Global Life Cycle Impact Assessments of Material Shifts

A.S.G. Andrae

Global Life Cycle Impact Assessments of Material Shifts

The Example of a Lead-free Electronics Industry

Springer

Dr. A.S.G. Andrae
Huawei Technologies Sweden AB
Box 54
16494 Kista
Sweden
anders.andrae@huawei.com

ISBN 978-1-4471-5781-6 ISBN 978-1-84882-661-8 (eBook)
DOI 10.1007/978-1-84882-661-8
Springer London Dordrecht Heidelberg New York

British Library Cataloguing in Publication Data
A catalogue record for this book is available from the British Library

Softcover re-print of the Hardcover 1st edition 2010

Cover design: eStudioCalamar, Figueres/Berlin

Printed on acid-free paper

Springer is part of Springer Science+Business Media (www.springer.com)

Preface

We live in a complex world in which the competition between people and nations are increasing. Moreover is mankind as an "inhabitant" not always treating its "house" the Earth with due respect and care.

For instance the current enhanced greenhouse effect has the potential to drastically change living conditions. Alarmingly, the effect could be underestimated due to a reflecting mechanism of particle pollution. Moreover, around half of the "original" forests have been removed to give space for arable land, inducing a positive feedback for increased global heating. Another serious concern is that the numbers of animal species are decreasing partly due to Mankind's destructive ways of life. Many species provide "ecosystem services" which could be enormously valuable [1]. Bee-aided pollination is a striking example of such services.

Undisputed research has demonstrated that the Earth system is more than the sum of its parts so it is not recommendable to model the environmental impacts in a purely positivistic manner. Moreover, the ranking of different environmental problems is currently a task for legislators who drive the environmental initiatives of individual actors. Signs are showing that legislators are trying to introduce monetary fees on environmental loadings beyond e.g., sulphur dioxide, where the carbon dioxide taxes and trading systems are the most obvious examples. Recently the Swedish government suggested increasing the tax on ammunition which contains lead (Pb) in order to promote the development of alternatives.

More and more governments seem to understand that the current global economy is not sustainable from an environmental viewpoint. In order to grasp what kinds of environmental aspects could be involved in the electronics industry, a definition of an environmental problem is needed. An environmental problem can be regarded as such, if these three criteria are fulfilled: first, an objective change in the environment is to take place; second, this change shall have to be human induced; and third, this change shall be regarded as a problem by society. The vast and tremendously multifaceted electronics industry is characterized by its long supply chains and thereby many indirect environmental impacts associated with its products.

The microelectronic products used in everyday life seem small and harmless, but quite a lot of material and energy is needed to produce electronic components, especially the integrated circuit (IC).

The life cycle schematically consists of three phases which are manufacturing followed by the use phase, and last the end-of-life. One important network of manufacturing processes is the one associated with the production of ICs. However, the final assembly of electronics does not give rise to tremendous global environmental problems, compared to the processes further upstream.

The cradle-to-gate usage of primary energy (PEU) is around 2,300 MJ/kg and the CO_2 equivalent intensity 150 kg/kg for printed board assemblies[1] (PBA). However, the PBAs differ dependent on the electronic device in which they are used. *According to estimations* for an advanced telecom PBA, the silicon-based ICs are near 30% and printed wiring board is around 40% of the PEU. The so-called high-purity chemicals used in silicon wafer processing could at least make up 5% of the PEU for a PBA from "cradle-to-gate". The solder/adhesive part of the PBA is below 1% of the CO_2 and PEU scores. It seems that the solder/adhesive component is mostly interesting from an environmental risk management perspective in the waste treatment phase, and not due to energy usage concerns.

During the use phase the electronic products use electricity and the environmental load is dependent on how this electricity has been obtained from secondary energies. According to the International Energy Agency, combustion of the fossil fuel coal is the most common method worldwide and coal makes up around 40% of the primary energies used to obtain useful electricity. Global electricity conversion is also associated with metal emissions such as Hg and Pb. It is likely that the electricity usage worldwide will significantly increase, as a result of increased use of electronics. This development is part of the westernization of the world emphasized by the economic growth in, e.g., China, India, and Brazil.

In the developed world, at the end of the electronic product life only a small share is currently recycled, mostly to get hold of the valuable metals. (In the developing part of the world the electronics recycling situation is even worse.) The rest is first incinerated and then deposited in landfills, or directly deposited. The export of used products, e.g., obsolete computers, from the developed countries to developing is significant and increasing [2]. The microelectronics business is challenged by a number of recent environmental regulations. It is likely that an EU Directive addressing the recycling of waste from electronic and electrical equipment, WEEE, will improve the recycling rate in Europe [3]. Heavy-metal emissions (e.g., Pb in solders) occur as a result of improper recycling and therefore the global flows of electronics waste from the industrialized world to the developing one is currently leading to occupational health problems [4].

According to the European Community Directive 2000/0159 C5-0487/2002 (RoHS) regarding limitation of the use of certain environmentally harmful materials, the Member States were obliged as of July 1 2006 to ensure that, among other materials, Pb is not used in soldering materials employed in electrical and electronic equipment to be sold on the market [5]. Nevertheless, Pb can still be used for solder alloys containing 85wt% Pb.

---

[1] Printed wiring board = printed circuit board = printed board = a board without components and interconnection materials

Actually, one of the most controversial issues that the electronics industry handles is the shift to Pb-free production. The Pb ban RoHS was proposed as Pb has been banned in other products, and also the legislators estimated the environmental risk, of Pb in electronics waste, as high for humans and biota. One reason for the legislators' worry was that in fact electronics recycling is not happening quickly enough. However, several actors have individually and together demanded exemptions due to cost and reliability concerns of Pb-free alternatives.

For the environmental scientist it should be rather straightforward to shed light on whether this ban of a specific Pb usage makes sense. It mainly depends on data available for the simulations and the uncertainty associated with those data. There exist a number of approaches to tackle the problem. One would be to test the following two hypotheses:

1. The phasing out of Pb from the electronics industry (mainly interconnection materials) is preferable seen from *different environmental cost perspectives*
2. Pb emissions will decrease (*in the same way as they did when Pb was phased out as additive to petrol*) if Pb is phased out from the electronics industry

In this book I present my own environmental life cycle assessment (LCA) based research containing data with which these specified hypotheses are tested. The book elaborates on the ability of LCAs to answer questions of global character. This covers certain conclusions which can be drawn from the perspective of LCA on the transitional Pb-free ban in the electronics industry. Special emphasis is given to consequential LCA, uncertainty analysis and the Japanese LIME impact assessment modeling system. The LCA approach has many well-known problems, but also many positive features, both of which I will account for in detail for the specific applications. Environmental impact predictions definitely belong to holistic science in contrast to analytic reductionism. This is because the environmental systems are complex and e.g., the LCA approach demands multidisciplinary thinking which in turn needs the knowledge and methodology from more than one discipline.

References

1. Pimentel D, Wilson C, McCullum C (1997) Economic and environmental benefits of biodiversity. BioScience 47:747–757
2. Williams E, Kahhat R, Allenby B et al (2008) Environmental, social, and economic implications of global reuse and recycling of personal computers. Environ Sci Technol 42:6446–6454
3. European Union (2003) Directive 2003/ 108/EC of the European Parliament and of the Council of 8 December 2003 amending Directive 2002/96/EC on waste electrical and electronic equipment (WEEE)
4. Coby SC, Wong SC, Wu et al (2007) Trace metal contamination of sediments in an e-waste processing village in China. Environ Pollut 145:434–442
5. European Union (2003) Directive 2002/95/EC of the European Parliament and of the Council of 27 January 2003 on the restriction of the use of certain hazardous substances in electrical and electronic equipment. Off J Eur Union, 13.2.2003:L37/19–23

Acknowledgments

I would like to express my gratitude to Prof. Norihiro Itsubo at Musashi University of Technology in Japan for giving me the opportunity to work in his group and study the broad and developing field of life cycle assessment applied to electronic materials, and for all the support, motivation, and opportunities he has given me, leading to the opportunity of writing this book. Also, many thanks to all my colleagues and the staff in the LCA methodology group at National Institute of Advanced Industrial Science&Technology (AIST) in Tsukuba, Japan, who supported my work for the last few years. I would like to thank Prof. Atsushi Inaba, Director of the Research Center for Life Cycle Assessment[2] at AIST for stressing the importance of evidence of one's research endeavours. I would also like to thank my former employers, especially Prof. Johan Liu, Chalmers University of Technology in Gothenburg, Sweden for their kindness to hire me, which gave me the inspiration to start my post-doctoral PhD studies in the first place. Thanks to Dr. Hiroshi Yamaguchi at the Musashi University of Technology, Yokohama, Japan, for helping me with data collection at Japanese companies and constructive criticism of my writings. Financial support has been provided by National Energy Development Organization, Japan Society for Promotion of Science, and Huawei Technologies and is gratefully acknowledged. My current employer Huawei Technologies Sweden AB is recognized for kindness in hiring me. I also thank Western Norway Research Institute in Sogndal, Norway and especially Dr. Otto Andersen for their professional interest in me. Finally, I am grateful to my beloved family and fiancée Yoko for their kind support.

Stockholm, Sweden *Anders Andrae*
July 2009

[2] The Research Center for LCA has since April 1, 2008 been included in the Research Institute of Science for Safety and Sustainability at AIST in Tsukuba.

Contents

Abbreviations

ALCA	Attributional Life Cycle Assessment
CLCA	Consequential Life Cycle Assessment
LCA	Life Cycle Assessment
LCI	Life Cycle Inventory
LCIA	Life Cycle Impact Assessment
ICA	Isotropically Conductive Adhesive
RoHS	Restriction of Hazardous Substances
WEEE	Waste of Electronic and Electrical Equipment
BD	Bio-Diversity
HH	Human Health
PP	Primary Productivity
SA	Social Assets
JPY	Japanese Yen
PBA	Printed Board Assembly
PWB	Printed Wiring Board
Sd	Total standard deviation as given by EMLCA
i	Environmental loading, emission or resource
A	Flow of products and materials
B	Flow of environmental loadings
g	Vector of environmental loadings
j	Environmental impact category
m	Damage impact category
n	Number of environmental loadings
q	Number of environmental impact categories
SC	Social cost
r	Number of damage categories
t	Damage factors
D_m	Overall indicator of the damage category
$D_{i(m)}$	Damage indicators
$I_{i(j)}$	Impact indicators
I_j	Overall indicator of the environmental impact category
k	Characterization factors

w_m	Normalization factors for overall damage indicators
w_j	Normalization factors for overall environmental impact indicators
f_u	Special unit process output
f	Demand vector
y	Price
k_d	Own-price elasticity of demand
k_s	Own-price elasticity of supply
x_d	Quantity demanded
x_s	Quantity supplied

List of Figures

List of Tables

Chapter 1
Introduction

Abstract The chapter presents an extensive background to the research subject of environmental impact analysis of interconnection materials, including conductive adhesives. The resource issues of tin and silver are briefly mentioned as well as the legislative Pb ban exemptions of high-melting-point Pb based solders. The differences between attributional LCA and consequential LCA are mentioned. The purpose of the book is clarified in the goal and scope section in which also the overall hypotheses are presented. The outline of the whole book is described.

1.1 Background

Industrial development in the microelectronic industry is rapid, and demands on new materials are getting more and more sophisticated. Solders and conductive adhesives are used by electronics manufacturers to provide electrical interconnections between electrical components and the printed wiring board (PWB). Since the 1940s, lead (Pb) has been used in electrical solders in solder alloys for its proven reliability in eutectic combination with tin (Sn) [1, 2].

For products, which can be classified as passive (i.e., energy is not required for their function), the major environmental impact throughout the life cycle is due to the production of materials or the disposal. Solders, conductive adhesives, and electronic component platings fall under the definition of passive materials [3]. Nevertheless the use phase of the solders and conductive adhesives is considered in the present book to be the stencil printing, component mounting, and wave/reflow processes, where energy is required to melt or cure the materials. Pb has been used in solders since the breakthrough of electronics [4]. The heavy metal Pb is under certain conditions harmful for humans, therefore it was banned from use in *petrol* in Canada in 1990 [5] in the U.S.A. via the Clean Air Act in 1996 [6] and in the European Union in 2002 [5]. Regarding *new water drinking systems* (plumbing solders) Pb was banned in 1988 in the U.S.A via the Lead Contamination Control Act [7] and for use in *house paint* in 1978 [8]. In summary, Pb is not banned totally in paint, drinking-water installations, and petrol in all nations of the world, but many countries have restrictions nevertheless. In Europe, an EU directive on Restrictions on the use of Hazardous Substances (RoHS) required the elimination of Pb in electronics, in all but special applications, by July 1, 2006 [9].

Chinese electronics production is widely affected by the European legislations [10]. In China itself, Order #39 from the Ministry of Information Industry restricts the use of Pb [11].

As electricity production is an important flow when comparing Pb-based solder to Pb-free alternatives, it will be important to understand the quantities and risks of Pb released from, e.g., the coal-burning power plant life cycle, compared to the quantities and risks of Pb emitted from solders. In 1994 Sullivan estimated that the Pb content of the coal needed to supply the power of an alarm clock over its life time was similar to the amount of Pb contained in the clock's solder joints [12]. *See Sect. 4.1.4 for more elaborations on electronic device life time Pb emissions.* In year 2000 a Printed Board Assembly (PBA) allowing Pb based interconnection typically contained 50 g Pb per m^2 [13]. Another study from year 2005 mentions 30 mg wave solder Pb/cm^2 and 7 mg reflow solder Pb, i.e., 3.7 g Pb per m^2 [14]. A measurement and calculation from year 2007 in the telecom industry gave 32 g Pb per m^2. In 2002 in China the Pb emissions to air from average electricity production were estimated 1.42 μg per kWh [15]. Pb compounds used in gasoline are still a concern due to atmospheric pollution [16]. Yasui argued that the reduction of risk due to exposure of Pb from solders will not change *in advanced countries* as a result of Pb-free legislation. The reasons are that the Pb usage for solders has decreased, and that the main exposure route of Pb is through burning different Pb-containing wastes [17].

Of the Pb present in European landfills in 2002, around 40wt% originated from consumer electronics [18]. In Sweden in 2007 around 12,000 tonnes of electronics waste was not treated properly. About 0.6wt% of the Swedish municipal household waste was electronics waste (e-waste), mainly light bulbs, small batteries, mp3 players, and keyboards, which should have been recycled. Globally, Sweden recycles the highest amount of e-waste per person, but perhaps as little as 10wt% of global e-waste is recycled [19]. In European Union Member States, the inflow of Pb in alloys to landfills and waste incineration is about 80wt% caused by Pb solders used in electronics [20].

Founded on a *hazard-based* approach, in contrast to a *risk-based* approach, this triggered a development of alternatives to Sn–Pb solders. Two arguments have been that recycling of e-waste would cause Pb-related health problems to workers, and that Pb could leach from landfills to the groundwater. Although Pb-free solders are not new, they have been little used to date [21]. Several researchers have tried to environmentally compare the Pb-free solders to Sn–Pb ones. Deubzer [22], Itsubo [23], and PRe' Consultants [14] on the one hand, and Turbini [24], Verhoef [25], Reuter [26], US EPA [27], and Ogunseitan [28], on the other, provide contradictory information concerning the environmental compatibility of Pb-free solders. The first group of authors is, judging from their final conclusions, somewhat neutral/positive toward Pb-free solders, while the second is rather negative. However, all authors seem to agree on the higher life cycle energy usage of Sn–Ag–Cu variants of Pb-free, compared to Sn–Pb soldering. Nevertheless, several authors argue that although the Pb-free shift will be disadvantageous from a global warming point of view, it might be advantageous from a toxic emissions point of view. Some of the environmental assessments are traditional environmental life cycle assessments (LCA), while others focus on specific aspects of the solder life cycle. LCA is a standardized method handling the quantification of potential environmental effects associated with, e.g., a technology such as interconnection materials. This book will present results, thoughts, and insights developed from Andrae's interconnection material LCA research [29–32] of which two studies compared silver- (Ag) and polymer-based electrically conductive adhesives (ECA) to Pb-based solders [30, 32]. Moreover, the resource issues, for example the possible scarcity of metals replacing Pb, could also be important for certain metals as in 2004 globally around 130,000 tonnes of solders were used by the electronics industry [22]. Rough estimations of global solder usage can be made by the following simple calculation:

```
Solder alloy consumption =
[Sn consumption × solder share of Sn consumption × e-
solder share of solders]/
[share of 63Sn-37Pb solder × 0.63 + density 63Sn-37Pb
solder/density 10Sn-90Pb solder × share of 10Sn-90Pb
solder × 0.1 + density 63Sn-37Pb solder/density Pb-
free solder × share Pb-free solders × 0.95]
```

The possible distribution between wave solder alloys and reflow solder paste is shown in Fig. 1.1.

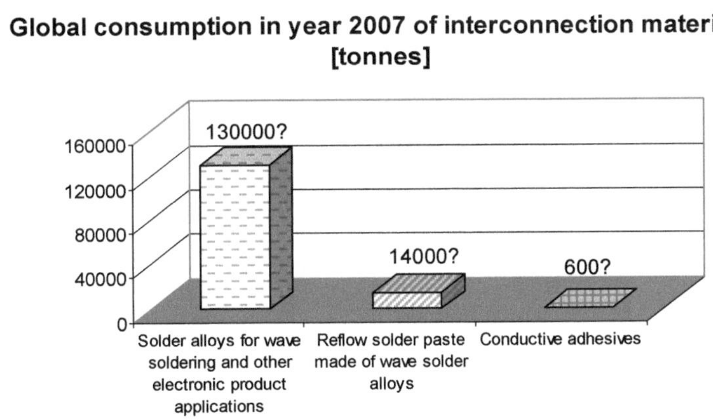

Fig. 1.1 Approximate global usage in tonnes of interconnection materials in 2007

According to recent estimations done by the International Tin Research Institute (ITRI) of Sn consumption, which is strongly correlated to electronic solder consumption, 362,000 tonnes was demanded globally in year 2006. The Ag consumption is not as strongly related to solder consumption but is also affected [33]. According to IPC the share of Pb-free solders was at 62% of the market in 2007. In 2006, according to ITRI, the share of solders of the total tin use was around 50%, and the US Geological Survey estimated the share of e-solders to about 65%. This would mean a solder alloy consumption for electronics of approximately 145,000 tonnes. Figure 1.2 shows the near trends of metal consumptions in interconnection materials.

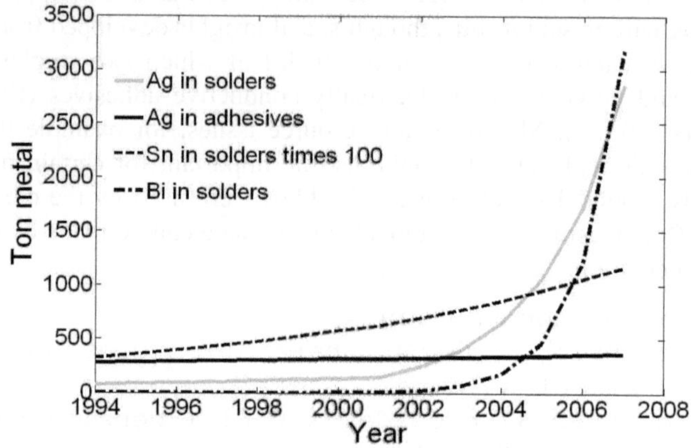

Fig. 1.2 Historical development of metal usages for solders and adhesives

The code in MATLAB® which generates Fig. 1.2:

```
x = [1994 1995 1996 1997 1998 1999 2000 2001 2002 2003
2004 2005 2006 2007];%years
yAgs = [77.52   85.272  93.7992 103.17912   113.497032
124.8467352 137.3314087 150 244.5   398.535 649.61205
1058.867642 1725.954256 2862]; %Ag usage in solders
yAga = [280 285.6    291.312 297.13824    303.0810048
309.1426249 315.3254774 320 326.4    332.928 339.58656
346.3782912 353.305857  360];%Ag usage in adhesives
ySns  =  [326  358.6        394.46      433.906   477.2966
525.02626          577.528886      626   694.86        771.2946
856.137006   950.3120767 1054.846405 1170] ; %Sn usage
in solders
yBis = [1   2   3   4   5   6   7   10  26.1    68.121
177.79581    464.0470641 1211.162837 3200] ; %Bi usage
in solders
pAgs = polyfit(x,yAgs,1);
pAga = polyfit(x,yAga,1);
pSns = polyfit(x,ySns,1);
pBis = polyfit(x,ySns,1);
plot(x,yAgs,'-', x,yAga,'-', x,ySns,'-',x,yBis,'-')
```

```
xlabel('Year','FontSize',30)
ylabel('Ton metal','FontSize',30)
title('Development of metal usages for solders and ad-
hesives between 1994 and 2007','FontSize',30)
legend('Ag in solders','Ag in adhesives','Sn in solders
times 100','Bi in solders')
grid off
```

It is probable that the usage will increase despite miniaturization trends, thereby implying considerable increases in the global Sn and Ag consumptions [34, 35].

In 2007 the USA was around 20% of the total global end market for electronics worth ~ $\$2.8 \times 10^{11}$ (micro, opto, computers, TVs, mobile phones, etc.) and electronic manufacturing services (mainly assembly of PBAs where solder is used) was around 40% of the value of electronics USA end market. In 2007 solder for electronics in the USA was a ~ $\$6 \times 10^{8}$ annual market. Solder was therefore around 0.2% of the USA end market for electronics and presumably a similar share of the ~ $\$1.4 \times 10^{12}$ global end market for electronics, or ~ $\$3 \times 10^{9}$. The solder value is probably higher now as the global market for electronics grows by 7–8% per year in combination with growing usage of the more expensive Pb-free solders.

Ag-based conductive adhesives currently play a small role in the interconnection materials market with around 5% of the *solder paste* market. However, should these adhesives be extensively used, the global Ag consumption would be noticeably affected.

In this book, there is a distinction between the traditional attributional (ALCA) and consequential LCA (CLCA). Attributional methodology for life cycle inventory analysis (LCI) aims at describing environmentally relevant physical flows to and from a life cycle and its subsystems. Ideally, it should include average data on each unit process within the life cycle. The attributional LCI (ALCI) model does not include unit processes other than those of the life cycle investigated, but significant cut-offs within the life cycle can be avoided through the use of environmentally expanded input–output tables [29]. Most previous LCAs resemble ALCA which is described in Sect. 4.3.

In contrast, the aim of consequential LCI (CLCI) methodology is to describe how the environmentally relevant physical flows to and from the technosphere will change in response to possible changes made within the life cycle.

A CLCI model includes unit processes that are *significantly affected* irrespective of whether they are within or outside the life cycle. Ideally, it should include marginal data on "bulk" production processes in the background system [36].

In a CLCI allocation is usually avoided by so-called system expansion [37]. A CLCI model can also include economic partial equilibrium models [38]. In Fig. 1.3 is shown a conceptual scheme of which processes could be included in a CLCI of a Pb-free solder.

Several authors have made similar distinctions between the two types of LCA methodologies, although most authors employ different terms to denote them [39–42]. CLCI is described in Sect. 4.4.

ALCA has been used to compare 63Sn–37Pb to several Pb-free material combinations using Sn in combination with Bi, Cu, Co, In, Sb, Zn. Schoenung et al. presented a review of the scientific status of the complex problem related to the industrial ecology of Pb-free electronics. For LCA, they concluded that performed case studies lack transparency and therefore could not with ease be validated. Moreover, the legislations and policies regarding electronics waste differ between the EU, USA and Japan [43].

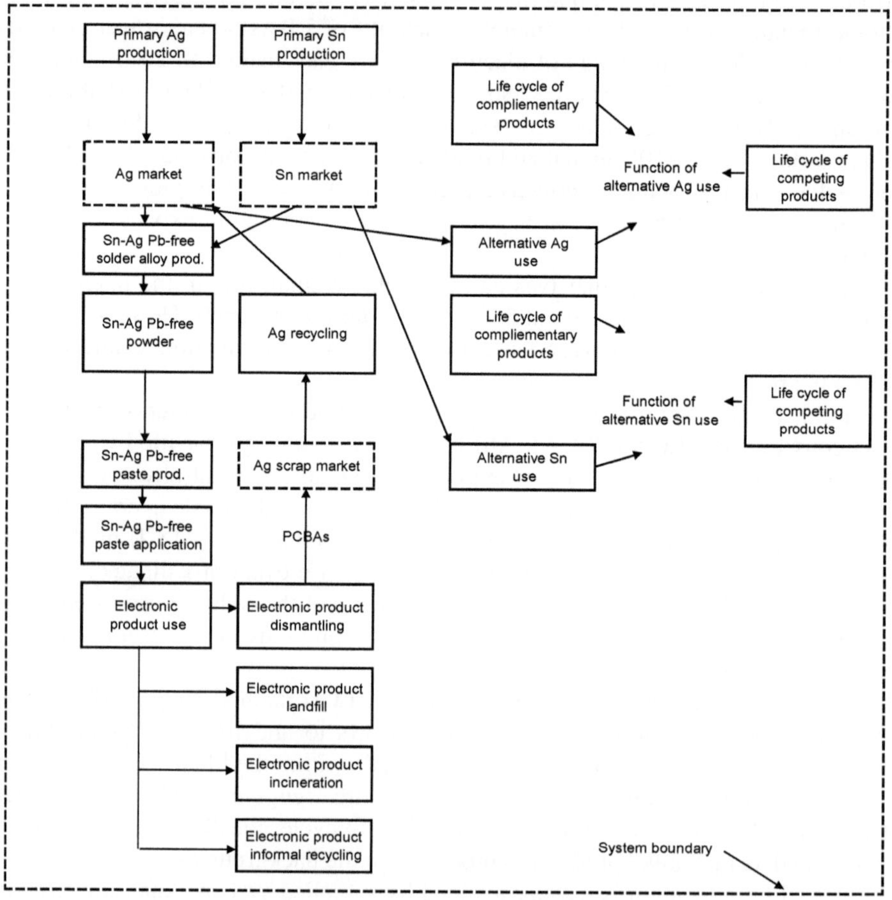

Fig. 1.3 A conceptual scheme of a CLCA of Pb-free solder paste

Life cycle impact assessment is an essential part of LCA. The environmental impacts, and trade-offs between them, can be estimated using methodologies such as the Life Cycle Impact Assessment Method based on Endpoint Modeling (LIME) [44]. LIME is described in some detail in Sect. 3.4.

Regarding interconnection materials, used in reflow soldering and curing processes to mount components onto the printed wiring boards, three main choices remain after the conventional Pb-based solders have been finally banned: Pb-free solders, ECAs or solutions which do not use interconnection materials. ECAs are becoming interesting as research seems to be near a breakthrough for some of the main issues for material properties and assembly. ECAs consist of a polymeric binder matrix, usually bisphenol-A type epoxy resin, and metal fillers where the metal usually is Ag, Au, Cu, or Ni.

Another application is bonding prior to wave soldering and surface mounting of electric components, for the latter isotropically conductive adhesives will be studied in the present research. Furthermore, different surface plating metals are used to protect the pads of the printed wiring boards and the component leadframes.

The analysis of environmental impact of high-temperature conductive adhesives compared to high-melting point (HMP) Sn–Pb solders and HMP Pb-free solders would be relevant avoid introducing a worse alternative for a long time. In the present book, HMP Pb-free solders will be briefly assessed as it is judged that conclusions can be drawn based on LCAs performed on low-melting-point (LMP) Pb-free solders.

1.2 Goal and Scope

The aim of the book is to give a detailed account on the life cycle inventory and impact assessment procedure for assessment of interconnection materials used in the electronics industry. The emphasis will be on uncertainty analysis, consequential approaches and LIME-based LCIA.

The overall hypotheses that this book addresses are:

1. The environmental costs associated with Pb-free and Pb-based interconnection materials can initially and meaningfully be estimated using life cycle assessment.
2. The consequential environmental cost of a certain shift from Pb-based interconnection material to Pb-free can be quantified using consequential LCA.

1.3 Outline of the Book

The technical research field of interconnection materials is described in Chap. 2 and LCA in Chap. 3. Chapter 3 includes detailed definitions of LCA and LCI as well as explanations of allocation and examples of general LCAs. Moreover the

LCIA method LIME is accounted for and additionally there is a review of other authors LCA research and case studies of interconnection materials.

In Chap. 4 the methodology specifically used in the present research is presented including LCA software, the consequential LCI approach and uncertainty analysis techniques. In Chap. 5 and 6 the results of performed research are discussed in appropriate sections. In Chap. 7 the overall results are discussed, whereas Chap. 8 extracts the conclusions and Chap. 9 provides an outlook for the future.

References

1. Gonser BW (1949) Tin and its alloys. Ind Eng Chem 41:2147–2149
2. Plumbridge WJ (1996) Solders in electronics. J Mater Sci 31:2501–14
3. Rydh CJ, Sun M (2005) Life cycle inventory data for materials grouped according to environmental and material properties. J Clean Prod 13:1258–1268
4. Li Y, Moon KS, Wong CP (2005) Electronics without lead. Science 308:1419–1420
5. Bollhöfer A, Rosman KJR (2001) Isotopic source signatures for atmospheric lead: the Northern Hemisphere. Geochim Cosmochim Acta 65: 1727–1740
6. Kovarik W (2005) Ethyl-leaded gasoline. Int J Occup Environ Health 2005:384–397
7. US Lead Contamination Control Act of 1988 PL 100-572 (amends Safe Drinking Water Act) [Nov. 1, 1988]
8. US Consumer Product Safety Commission of 1978 (16 Code of Federal Regulations CFR 1303).
9. European Union (2003) Directive 2002/95/EC of the European Parliament and of the Council of 27 January 2003 on the restriction of the use of certain hazardous substances in electrical and electronic equipment. J Eur Union, 13.2.2003:L37/19–23
10. Tong X (2007) Diffusion of lead-free soldering in electronics industry in China. Chin Popul Res Environ 17: 66–71
11. Ministry of Industry and Information Technology of the People's Republic of China (2004) MII Order #39
12. Sullivan CR (1994) Energy use: the most important impact of electronic equipment? Proc. IEEE Int Symp Electron Environ, 2–4 May, San Francisco, CA, 261–266
13. Five Winds International (2001) Toxic and hazardous materials in electronics, Environment Canada, 77
14. PRe' Consultants B.V. (2005) LCA comparison of alternative soldering techniques. Table 13, 17
15. Di X, Nie Z, Yuan B et al (2007) Life cycle inventory for electricity generation in China. Int J LCA 12:217–224
16. Pimentel D, Cooperstein S, Randell H et al (2007) Ecology of increasing diseases: population growth and environmental degradation. Hum Ecol 35:653–668
17. Yasui I (2007) Transition of environmental issues – fundamental criteria for LC engineering. Proc 14th CIRP Conf Life Cycle Eng – Advances in life cycle engineering for sustainable manufacturing businesses, Springer-Verlag London
18. Verhoef EV, Reuter MA, Scholte A (2003) A dynamic LCA model for assessing the impact of lead free solder. Proc Metallurgical and Materials Processing Principles and Technologies (Yazawa International Symposium) – Vol. 2: High-Temperature Metal Production, 2–6 March, San Diego, CA., 606
19. Ladou J, Lovegrove J (2008) Export of electronics equipment waste. Int J Occup Environ Health 14:1–10

20. Tukker A, Buijst H, van Oers L et al (2001) Risk to health and the environment related to the use of lead in products. TNO Report STB-01-39, 65
21. Andersson C (2007) Reliability, fatigue, and mechanical characterisation of lead-free solders for electronic packaging applications. PhD thesis, Chalmers University of Technology, Gothenburg, Sweden.
22. Deubzer O (2007) Explorative study into the sustainable use and substitution of soldering metals in electronics. PhD thesis, Delft University, Delft, the Netherlands
23. Itsubo N, Noh J, Inaba A (2004) Life cycle impact assessment of lead-free solder considering the trade-off relationship between toxic impact and other impact categories. Mater Trans 45:3194–3200
24. Turbini LJ, Munie GC, Bernier D et al (2001) Examining the environmental impact of lead-free soldering alternatives. IEEE Trans Electron Packag Manuf 24:4–9
25. Verhoef EV, Dijkema GPJ, Reuter MA (2004) Process knowledge, system dynamics and metal ecology. J Indust Ecol 8:23–43
26. Reuter MA, Verhoef EV (2004) A dynamic model for the assessment of the replacement of lead in solder. J Electron Mater 33: 1567–1580
27. US Environmental Protection Agency (2005) Solders in electronics: a life cycle assessment; EPA, Cincinnati, OH
28. Ogunseitan OA (2007) Public health and environmental benefits of adopting lead-free solders. JOM 59:12–17
29. Ekvall T, Andrae ASG (2006) Attributional and consequential environmental assessment of the shift to lead-free solders. Int J LCA 11:344–353
30. Andrae ASG, Itsubo N, Yamaguchi H et al (2007) Screening life cycle assessment of silver-based conductive adhesive vs. lead-based solder and plating materials. Mater Trans 48:2212–2218
31. Andrae ASG, Itsubo N, Inaba A (2007) Global environmental impact assessment of the Pb-free shift. Sold Surf Mount Technol 19:18–28
32. Andrae ASG, Itsubo N, Yamaguchi H et al (2008) Life cycle assessment of high-temperature Japanese conductive adhesives. Env Sci Tech 42: 3084–3089
33. Nakamura S, Murakami S, Nakajima K et al (2008) Hybrid input-output approach to metal production and its application to the introduction of lead-free solders. Env Sci Tech 42:3843–3848
34. Andrae ASG, Itsubo N, Yamaguchi H et al (2007) Conductive adhesives vs. solder paste: a comparative life cycle based screening. Proc 14th CIRP Conf Life Cycle Eng – Advances in life cycle engineering for sustainable manufacturing businesses, Springer-Verlag London:285–290
35. Eckelman MJ, Graedel TE (2007) Silver emissions and their environmental impacts: a multilevel assessment. Environ Sci Technol 41:6283–6289
36. Ekvall T (1999) System expansion and allocation in life cycle assessment – with implications for wastepaper management. Chalmers University of Technology, Gothenburg, Sweden
37. Tillman AM (2000) Significance of decision-making for LCA methodology. Environ Impact Assess Rev 20:113–123
38. Ekvall T, Weidema BP (2004) System boundaries and input data in consequential life cycle inventory analysis. Int J LCA 9:161–171
39. Weidema BP (1993) Development of a method for product life cycle assessment with special references to food products. Technical University of Denmark, Lyngby, Denmark
40. Udo de Haes H, Wrisberg N (1997) Life cycle assessment: state-of-the-art and research priorities. Ecomed Publishers, Landsberg, Germany, 59–88
41. Heijungs R (1997) Economic drama and the environmental stage-formal derivation of algorithmic tools for environmental analysis and decision support from a unified epistemological principle. Leiden University, The Netherlands
42. Hofstetter P (1998) Perspectives in life cycle impact assessment – a structured approach to combine models of the technosphere, ecosphere and valuesphere. Kluwer, Dordrecht

43. Schoenung JM, Ogunseitan OA, Saphores JDM et al (2005) Adopting lead-free electronics: policy differences and knowledge gaps. J Indust Ecol 8:59–86
44. Itsubo N, Sakagami M, Washida T et al (2004) Weighting across safeguard subjects for LCIA through the application of conjoint analysis. Int J LCA 9:196–205

Chapter 2
Interconnection Materials – Technical Research Status

Abstract This chapter briefly summarizes some of the history and the technical research status of the wide field of interconnection materials and associated plating materials. The overview of different materials sets the scope and perspective of the following environmental life cycle assessments discussed in Chaps. 5 and 6. Solder pastes based on nanosized metal spheres are introduced. Research exploring ways to produce electronics without interconnection materials is considered. Extensive technical-based research has been done by several groups and the present summary is by no means comprehensive.

2.1 Solders

Solders are by far the most common of interconnection materials and the solders used in wave soldering for hole mounting are by mass and volume used more than solder pastes in reflow soldering.

An even smaller share is occupied by solder alloys for bumping of microchips. In the US, John R. Barnes is compiling a massive bibliography for "designing lead-free, RoHS-compliant, and WEEE-compliant electronics".

By July 2009 Barnes concluded that 254 books and 216 PhD/MSc theses and thousands of different kinds of articles had been published about interconnection materials and closely related topics [1]. Barnes compilation manifests that monumental amounts of ideas and research have been produced within the multidisciplinary area of interconnection, meaning that it is an intriguing area for study.

A genuine understanding of soldering demands several research disciplines to come together.

For Sn–Pb solders the reliability is fairly well understood but the Pb-free solders are so far much less investigated. The correlation between scientific disciplines involved in soldering is shown in Fig. 2.1.

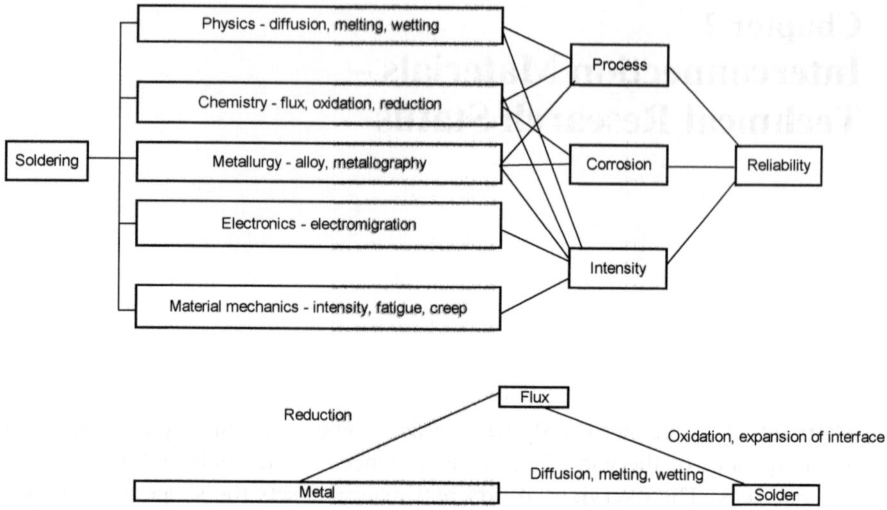

Fig. 2.1 The correlation of scientific disciplines determining the reliability of solder joints

2.1.1 Pb-based Low-melting-point Solders

The most common and oldest solder is the eutectic 63Sn–37Pb solder. It has been in use since the birth of the electronics industry. 63Sn–37Pb has a relatively low melting point, good wetting behavior, high electrical conductivity and can be used in hierarchical soldering. It is relatively common to use Sn–Pb solders having up to 2wt% Ag.

2.1.2 Pb-based High-melting-point Solders

In Europe, an EU directive on restrictions on the use of hazardous substances (RoHS) required the elimination of Pb in LMP solders, by July 1 2006.

However, the Pb-based HMP solders are exempted from the Pb ban of the RoHS but this is not expected of a forthcoming update of the Chinese version of RoHS [2].

The reasons behind the exemption are the relatively small amounts used of these kinds of solders, around 10wt% of the solder alloy production, and the difficulties in finding a technically viable substitution.

HMP Pb-based paste is applied before the main PWB is processed, using LMP solder/ECA and temperature sensitive components which would otherwise be damaged by the processing temperature required by HMP Pb-solders.

The high-melting-point solders are used in at least three areas: 1) as non-collapsible spheres (solder balls) in ball grid array joints between low temperature co-fired ceramic multichip modules and the mother PWB [3], 2) in ball grid array (BGA) joints between BGA components and the mother PWB, and 3) as solder within power device packages where the die is soldered to a leadframe. Figure 2.2 shows a hole mounted integrated circuit package where Level 1 is where high-melting-point solder is needed and at Level 2 low-melting point.

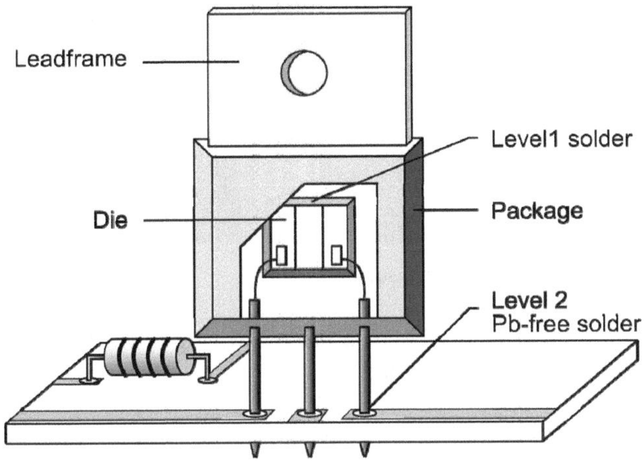

Fig. 2.2 Different usage of high- and low-melting solder in hole-mounted integrated circuit

BGA is a kind of semiconductor packaging technique. The solder manufacturers Indium and Kester call the Level 1 solders "die attach solders". These solders can also be used as alloys in wave soldering and reflow soldering as solder paste. The composition usually is usually more than 85wt% Pb and the rest Sn.

More than 900 combinations of ternary and quaternary Pb-free alloys could theoretically replace high Pb alloys but no "drop-in" alloy has been identified [4]. Section 2.1.4 mentions the latest research in high temperature Pb-free.

2.1.3 Pb-free Low-melting-point Solders

For the last decade Pb-free solders have been heavily researched. A wide variety of material combinations have been tested for many kinds of hypotheses. Consumer electronic applications are the primary target of the Pb-free solders. However, they are not just drop-in substitutes for traditionally used Pb solders due to solder melt temperature, processing temperature, wettability, mechanical and thermo-mechanical fatigue (TMF) behaviors, etc. [5]. An advantage of Pb-free solders is the higher service temperature capability provided to the solder joints. Figure 2.3 shows resistors through-hole mounted using Pb-free solder.

Fig. 2.3 Resistors mounted using Pb-free solder

Figure 2.4 shows the under side of a printed wiring board where an integrated circuit has been through-hole mounted.

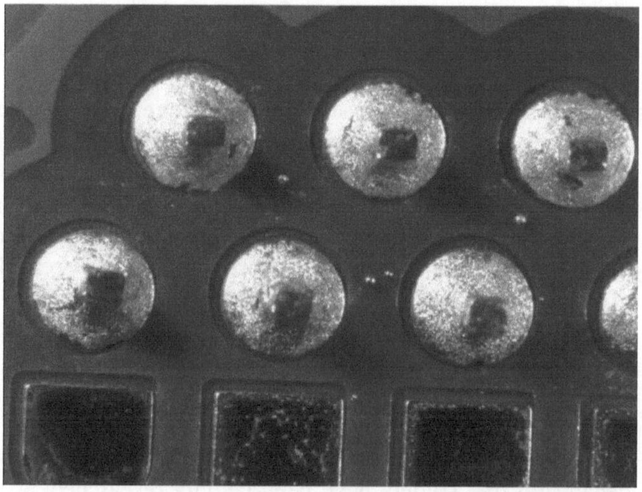

Fig. 2.4 Solder joints of an integrated circuit shown from the below of a printed wiring board

However, at the high temperatures required to reflow Pb-free alloys, moisture absorbed into the package can result in delamination and failure. On the other hand, the Pb-free alloys in general seem to improve the thermomechanical fatigue and electromigration performance, but then again induce risks for mechanical shock and whiskering.

For LMP solders, Sn will be the base but the Ag fraction might have competition from other materials such as Cu, Zn, Co, Bi, and rare earth metals [6–8].

The low-melting point Pb-free solders, e.g., the Sn–Zn system, have advantages compared to the Sn–Ag–Cu system. These are especially melting points near to the traditional Sn–Pb solder, low energy depletion, and no need for new surface mount technology (SMT) line.

Some low-temperature solders having melting point below 150 ºC can be used instead of conductive adhesives.

These are, e.g., 42Sn–58Bi, Sn50–In50, and 26Sn–53Bi–26Cd [9]. The latter is doubtful due to Cd toxicity.

2.1.4 Pb-free High-melting-point Solders

Pb-free high-melting-point solders are rather scarcely researched at the moment. The reason is that Pb-based high-melting-point solders are exempted from European Union Pb-free legislations.

Anyway, the main alternatives are 80Au–20Sn solders, Cu–Sn diffusion bonding, and liquid solders.

Liquid solders, which are at the experimental stage, are encapsulated solder joints with Bi and Sb containing solders, liquefying at higher temperatures.

Cu–Sn interdiffusion bonding is currently used e.g., in wafer stacking or chip to wafer assembly for System-In-a-Package.

80Au–20Sn solders are thought to be used only in niche applications due to the high cost of Au.

For high-temperature situations the bases instead of Pb could instead be Zn or Bi [10, 11].

2.1.5 Pb-free Solder Paste Based on Nanoparticles

One of the drawbacks with Pb-free solders similar to 96.5Sn–3.5Ag is their relatively high melting point being around 30 ºC higher than for 63Sn–37Pb eutectic solder.

The higher melting point brings about higher energy usage and also package reliability concerns such as substrate warpage and thermal stress [12].

When the sizes of the metal alloy spheres are reduced to nanometer size, the melting point can be reduced to from 217 to 213.9 ºC [13].

2.2 Conductive Adhesives

What is an electrically conductive adhesive (ECA)? One way of describing it is as a kind of polymeric solder that has been investigated by the electronics industry.

ECAs consist of a polymeric binder matrix (about 50vol% or 20wt%), usually bisphenol-A type epoxy resin, and metal fillers where the metal usually is Ag, Au, Cu, or Ni. It is also possible to have metal coated polymer spheres within the polymer matrix [14].

Li et al. suggest that ECAs generally are more environmentally friendly than solders as both Pb and flux cleaning are eliminated, and fewer overall processing steps are required [15]. Moreover, lower curing temperatures are thought to offer reduced energy use. Nevertheless, ECAs are not drop-in replacements for conventional solders due to different material properties and reliability issues such as low conductivity, unstable contact resistance, low joint strength, and Ag migration. Moreover, connected to the shift to Pb-free solders, ECAs need to be able to withstand considerably greater levels of thermal and chemical stress in order to prevent chip loss and component movement.

The wide usage of ECAs has so far only been for niche applications, especially to attach silicon dies to the microcircuit plastic capsules. ECAs are already used for attaching so called flip-chips on PWBs and for fastening parts of display units in, e.g., mobile phones, but have been considered too unreliable to be applied as full-scale interconnection materials [16].

Another application is bonding prior to wave soldering and surface mounting of electric components, for the latter isotropically conductive adhesives will be studied in the present research.

Furthermore, different surface plating metals are used to protect the pads of the printed wiring boards and the components leadframes.

Depending on the share of metal particles, ECAs are usually divided into three groups; isotropically conductive adhesives (ICA), anisotropically conductive adhesives (ACA), and non conductive adhesives (NCA) [17]. Moreover, ECA pastes can be used in standard processes, such as stencil printing and curing in conventional reflow ovens [18].

The main driver for development of ECAs was environmental hazard concerns about the Pb content of solder.

Much more research has been done for Sn–Ag–Cu solders than ECAs [19]. Some of the technical drives for ECAs have been the possibility to be used together with non-solderable materials such as chip-on-glass or surface mounted devices-on-polyester materials as well as the ECA lower sensitivity to thermomechnical stresses due to higher flexibility than solder [20].

Moreover, it is possible that ECA could replace solder pastes for certain applications. Further for ECAs, as, e.g., the oxidation problems are overcome, materials such as Cu, Ni, and AlN can replace Ag [21–23].

2.2.1 *Isotropically Conductive Adhesives (ICAs)*

In an ICA, the electrical contact is obtained through a network of contact points between individual particles [24]. The volume fraction of conductive particles is around 25–30%. ICA research data has emphasized failure, rather than reliability [19]. The particles consist of pure Ag, Au-plated Ag, Ag-plated Cu, or Ag-plated Sn but even more materials have been tested.

2.2.2 *Anisotropically Conductive Adhesives (ACAs)*

Conventional ACA is an adhesive consisting of conductive particles dispersed in an adhesive matrix. The volume fraction of conductive particles is around 5–10 %. These particles can be pure metals such as gold, silver, or nickel, or metal-coated particles (see Sect. 2.2.2.1) with plastic or glass cores. Palladium-plated Ag particles are less prone to Ag migration than pure Ag particles.

The volume fraction of particles is well below the percolation threshold, with the particles typically ranging from 3 to 15 μm in diameter. Typical for ACAs is that there is no contact between individual particles, the electric contact is instead achieved perpendicular to the film. ACAs are available in paste or film form (ACF) [24]. The reliability of an ACA is chiefly decided by the thermomechanical properties of the conductive particles. Figure 2.5 shows a sketch of flip-chip interconnection made possible by ACA.

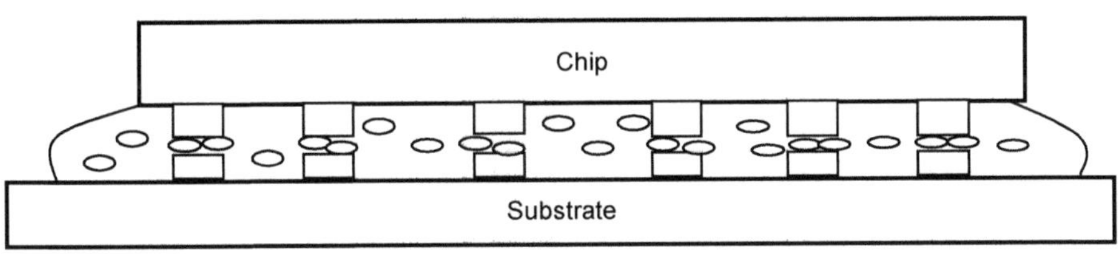

Fig. 2.5 ACA role in flip-chip interconnection

2.2.2.1 Metal-coated Polymer Spheres

Metal plated microspheres to be used in ACA, or as replacement for solder balls within ball grid array, or chip scale package components, are under development. The first application is shown in Fig. 2.6.

One alternative is a divinylbenzene polymer core plated with Ni and Au [25]. Acrylates and styrenes also work as core material as the cores can be made using the Ugelstad process after which the polymer core is then step by step covered with metals.

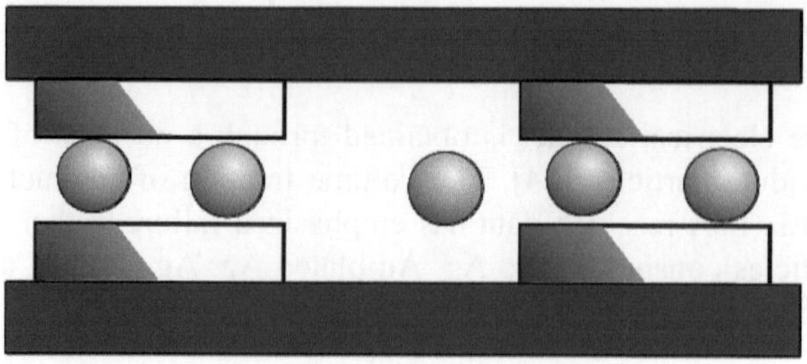

Fig. 2.6 Metal-plated polymer spheres used in ACA

2.2.3 Non-conductive Adhesives (NCA)

NCAs are characterized by zero volume percent added particles. The purpose of an NCA is to provide a mechanical connection between the contacts of a microchip and corresponding contacts on the substrate. The NCA bonding method relies upon direct electrical contact between the two conductor surfaces and the contact points are responsible for the transport of electrical current.

The epoxy filled cavities supply the adhesive forces needed to keep the materials together. After the connections are made, shrinkage in the cured adhesive and the mechanical properties of the involved materials will be responsible for the compressive force needed to maintain the electrical contacts [24].

2.2.4 Conductive Adhesives Based on Nanoparticles

Professor Johan Liu's research group at Chalmers University in Sweden is exploring the nanosize effect in flip-chip interconnections. It is done via ACAs containing nanoparticles of Pb-free solder alloys as fillers [26].

2.3 Solder-free Alternatives

Being rather "provocative" but nevertheless tremendously interesting and having inherently large implications for the whole Electronic Manufacturing Service industry, Californian company Verdant Electronics in 2007 boldly proposed a totally solder-free new production concept [27, 28]. However, it remains to be seen if solder-free alternatives will be successful in the interconnection market place.

2.4 Plating Materials and Metal Spheres for Ball Grid Arrays

The amount of plating materials, e.g., Pb-based alloys, involved within the components is less significant compared to the materials used to connect them to them the printed wiring board. For example a thin quad flad pack (TQFP) component having 176 connections by Xilinx and 0.5 mg solder paste per connection, the amount of 63Sn–37Pb solder paste to 85Sn–15Pb external plating would be 0.088 g to 0.016 g. The Pb content is 2 to 1. As for BGAs and CSPs, the solder balls are attached to the motherboard using traditional solders [29].

It is possible that some, but not all, BGA components need more solder from the solder balls and solder paste than a comparable surface-mounted QFP packages needs solder paste.

A metal-based BGA352 by Xilinx needs more than 0.8 g solder, 0.7 g 63Sn–37Pb solder for solder balls and around 352 (number of balls) × 0.8 (80% of area covered) × 0.08^2 (area, cm^2) × 0.015 (solder height, cm) × 4.7 (density solder paste, g/cm^3) = 0.1 g for 63Sn–37Pb solder paste. TQFP176 only needs around 0.2 g for solder paste and plating.

The silicon die within these packages weighs 0.028 g but the Input/Output connections are more for the BGA.

Then on the other hand a plastic-based BGA596 from Philips needs 0.94 mg for solder balls and the comparable Philips QFP64 needs 32 mg. The silicon die inside these packages weigh 20 mg.

Packages are very different but, in order to be comparable, they have to 1) fulfill the same function, 2) divert heat equally and 3) withstand environmental pressures, e.g., moisture, equally efficient. Bearing these criteria in mind, there would be room for developing a road-map for the most eco-friendly packages.

Figure 2.7 shows how an identical die is used in different packaging technologies.

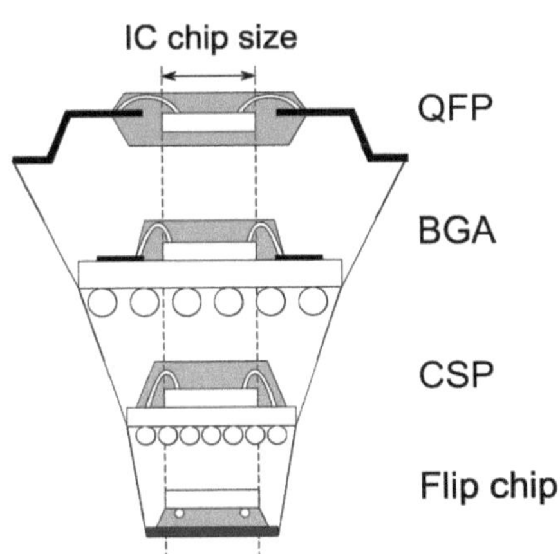

Fig. 2.7 The "footprint" of different packaging technologies

2.4.1 Component Terminal Platings

Two of the most common component terminal platings are Sn–Pb alloys and Ag on Pd. 85Sn–15Pb has been reported by microcircuit manufacturers such as Xilinx. Also Au/Pd/Ni or only Sn is used [30].

2.4.2 Printed Wiring Board Pad Platings

For PWB pad platings, bare Cu pads, Sn–Pb alloys or Au-plated boards are common. Also Au/Pd/Ni or only Sn is used [30].

2.4.3 Metal Spheres for Ball Grid Arrays

The most common materials are 63Sn–37Pb and increasingly 96.5Sn–3Ag–0.5Cu [31–34].

2.5 Perspectives

Compatibilityfree with Sn–Pb with Sn–Pb alloys is an important criterion of a Pb-free alloy, as there are many areas of possible alloy inter-mixes which imply different degrees of reliability. Huang and Lee found that Sn–Pb balled/plated BGAs, CSPs, QFPs, or TSOPs soldered with Pb-free solders showed serious board level reliability risks [35]

Moreover, the replacement of high-melting-point Pb-based solders could prove to be somewhat difficult. The reason is that the minimum acceptable melting point for the "die attach solder" will have to increase as reflow temperatures of board level Pb-free solders could reach as high as 270°C [36]. Possibly high-temperature conductive adhesive technology will solve the aforementioned problem as these adhesives can withstand high temperatures, but at the same time do not require a high reflow temperature [37].

References

1. http://www.dbicorporation.com/rohsbib.htm
2. McCluskey FP, Dasha M, Wanga F et al (2006) Reliability of high temperature solder alternatives. Microelectron Reliab 46:1910–1914

3. Nousiainen O, Putaala J, Kangasvieri T et al (2006) Metallurgical reactions in composite 90Pb10Sn/lead-free solder joints and their effect on reliability of LTCC/PWB assembly. J Electron Mater 35:1857–1865
4. Lalena JN, Dean NF, Weiser MW (2002) Experimental investigation of Ge-doped Bi–11Ag as a new Pb-free solder alloy for power die attachment. J Electron Mater 31:1244–1249
5. Guo F (2007) Composite lead-free electronic solders. J Mater Sci: Mater Electron 18:129–145
6. Andersson C, Sun P, Liu J (2008) Tensile properties and microstructural characterization of Sn–0.7Cu–0.4Co Bulk Solder Alloy for Electronics Applications. J Alloy Compounds 457:97–105
7. Suganuma K, Kim KS (2007) Sn–Zn low temperature solder. J Mater Sci – Mater Electron 18:121–127
8. Wu CML, Wong YW (2007) Rare-earth additions to lead-free electronic solders. J Mater Sci – Mater in Electronics 18:77–91
9. Chriastelova J, Ozvold M (2008). Properties of solders with low melting point. J Alloy Compounds 457:323–328
10. Lee JE, Kim KS, Suganuma K et al (2007) Thermal properties and phase stability of Zn–Sn and Zn–In alloys as high temperature lead-free solder. Mater Trans 48:584–593
11. Takaku Y, Ohnuma I, Kainuma, R et al (2006) Development of Bi-base high-temperature Pb-free solders with second-phase dispersion: Thermodynamic calculation, microstructure, and interfacial reaction. J Electron Mater 2006:(35):1926–1932
12. Jiang H, Moon KS, Hua F et al (2007) Synthesis and thermal and wetting properties of tin/silver alloy nanoparticles for low melting point lead-free Solders. Chem Mater 19:4482–4485
13. Zou C, Gao Y, Yang B et al (2009) Nanoparticles of the lead-free solder alloy Sn–3.0Ag–0.5Cu with large melting temperature depression. J Electron Mater 38:351–355
14. He JY, Zhang ZL, Kristiansen H (2007) Mechanical properties polymer particles for adhesives Int J Mater Res 98:389–392
15. Li Y, Wong CP (2006) Recent advances of conductive adhesives as a lead-free alternative in electronic packaging: materials, processing, reliability and applications. Mater Sci Eng R 51:1–35
16. Yim MJ, Paik KW (2006) Recent advances on anisotropic conductive adhesives (ACAs) for flat, panel displays and semiconductor packaging applications. Int J Adhes Adhes 26:304–313
17. Morris JE, Liu J (2007) Electrically conductive adhesives – a research status review. In: micro- and opto-electronic materials and structures: physics, mechanics, design, reliability, packaging, 1st edn. Springer, New York
18. Mo Z (2005) Reliability and applications of adhesives for microsystem packaging. PhD thesis, Chalmers University of Technology, Gothenburg, Sweden
19. Morris JE (2007) Isotropic conductive adhesives: Future trends, possibilities, and risks. Microelectron Reliab 47: 328–330
20. Jagt JC, Beris JM, Lijten GFCM (1995) Electrically conductive adhesives: a prospective alternative for SMD soldering? IEEE Trans Comp Packag Manuf B 18:292–298
21. Zhao H, Liang T, Liu B (2007) Synthesis and properties of copper conductive adhesives modified by SiO2 nanoparticles. Int J Adhes Adhes 27:429–433
22. Goh GF, Yu H, Yong SS et al (2005) Synthesis and cure kinetics of isotropic conductive adhesives comprising sub-micrometer sized nickel particles. Mater Sci and Eng B 117:153–158
23. Yoshikawa J, Katsuda Y, Yamada N et al (2006) Electrically conductive aluminum nitride ceramics containing in-situ synthesized boron carbonitride. Key Eng Mater 317–318:653–656
24. Kristiansen H, Liu J (1998) Overview of conductive adhesive interconnection technologies for LCD's. IEEE Trans Comp Packag Manuf A 21:208–214

25. He JY, Zhang ZL, Midttun M et al (2008) Size effect on mechanical properties of micron-sized PS–DVB polymer particles. Polymer 49:3993–3999
26. Verma SC, Guan W, Andersson C et al (2006) Flip-chip interconnection using anisotropic conductive adhesive with lead-free nano-solder particles. IEEE Electron Systemintegration Technology Conf, Sept. 5–7, Dresden, Germany
27. Fjelstad J (2007) The Occam process: a new paradigm of promise for printed circuit manufacturing. CircuiTree 20:52–54
28. Fjelstad J (2008) Environmentally friendly assembly of robust electronics without solder. Circuit World 34: 27–33
29. Cohn C (2004) Packaging and interconnection of integrated circuits. In: Electronics packaging and interconnection handbook, Ed. Harper C, 1st edn. McGraw-Hill, New York
30. Andrae ASG, Itsubo N, Yamaguchi H et al (2007) Screening life cycle assessment of silver-based conductive adhesive vs. lead-based solder and plating materials. Mater Trans 48:2212–2218
31. Alam OA, Chan YC, Tu KN (2005) Effect of 0.5 wt % Cu in Sn–3.5%Ag solder balls on the solid state interfacial reaction with Au/Ni/ Cu bond pads for ball grid array (BGA) applications. Chem Mater 17:2223–2236
32. Shih PC, Lin KL (2007) Correlation between interfacial microstructure and shear behavior of Sn–Ag–Cu solder ball joined with Sn–Zn–Bi paste. J Mater Sci 42: 2574–2581
33. Liang J, Dariavach N, Callahan P et al (2007) Effects of thermal history on the intermetallic growth and mechanical strength of Pb-free and Sn–Pb BGA solder balls. Sold Surf Mount Technol 19:4–14
34. Choubey A, Osterman M, Pecht M (2008) Microstructure and intermetallic formation in SnAgCuBGA components attached with SnPb solder under isothermal aging. IEEE Trans Device Mater Reliab 8:160–167
35. Huang MK, Lee C (2008) Board level reliability of Pb-free designs of BGAs, CSPs, QFPs and TSOPs. Sold Surf Mount Technol 20:18–25
36. Lalena JN, Dean NF, Weiser MW (2002) Die attach solder design – calculation of phase diagrams for lead-free power IC die attach. Adv Packag, February 2002.
37. Andrae ASG, Itsubo N, Yamaguchi H et al (2008): Life cycle assessment of high-temperature Japanese conductive adhesives. Environ Sci Technol 42:3084–3089

Chapter 3
Environmental Life Cycle Assessment from a LIME Perspective

Abstract This brief chapter describes the method and in more depth the LIME method before arriving at the need for the present research. The crucial treatment of allocation, system boundary, and uncertainty estimations within LCA as well as applications of the tool is brought to attention. The broad topic of environmental cost quantification is discussed, bridging to the LIME section. The most important work in the environmental assessment of interconnection materials is reflected upon including metals, alloys, solders, conductive adhesives, and platings. Finally, the research need is clarified by asking eight questions based on the two hypotheses in Chap. 1.

It is generally difficult, and sometimes confusing, to quantify the environmental consequences associated with materials, processes and products. Difficulties are, for example, the determination of environmental effects associated with the objects of comparison, the nearly impossible task of comparing different environmental effects, and the amount of data needed to compare relatively simple products.

To further complicate evaluations, data needed for the analysis is also often classified or inaccessible.

On top of this, it is hard to delimit the environmental load analysis. The product life cycle is schematically shown in Fig. 3.1.

Nevertheless, there presently exist a number of methods and adherent tools for environmental assessment, such as life cycle assessment (LCA) and ecological footprint [1], all intended to indicate whether alternative A is better than alternative B or alternative C… or even alternative N.

The above mentioned methodologies have advantages for environmental experts such as:

- hot-spot finding within a life cycle;
- scenario analysis with the help of software;
- building-up of relational databases which can be reused in product development; and
- making trade-offs between environmental parameters in eco-design.

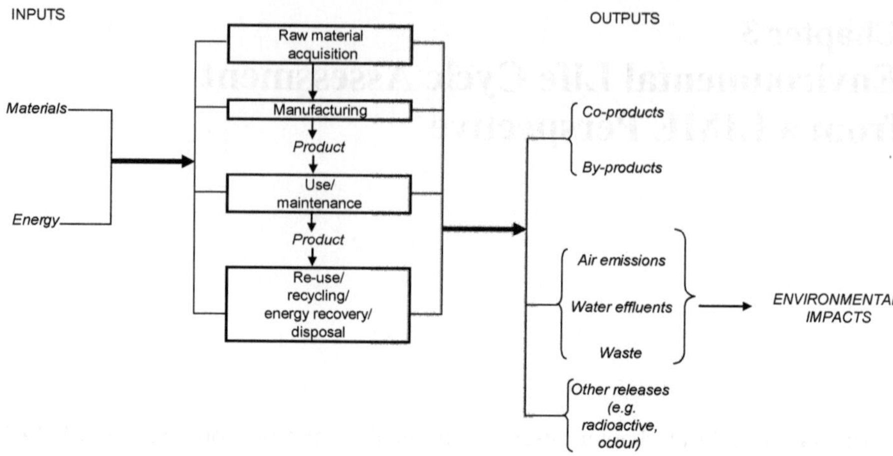

Fig. 3.1 The product life cycle which the LCA attempts to model

3.1 Definition of LCA and LCI

Life cycle assessment is a standardized methodological framework for estimating and assessing some of the possible environmental impacts attributable to the life cycle of a product or technology, such as climate change, stratospheric ozone depletion, tropospheric ozone (smog) creation, eutrophication, acidification, toxicological stress on human health and ecosystems, the depletion of resources, water use, land use, noise-and others [2].

The inventory flows, from either an attributional or consequential LCI, are classified according to which environmental impacts, e.g., global warming, they could cause. After the classification the flows are characterized according to their relative importance for each environmental impact indicator.

LCA generates useful information about which product has a more environmentally competitive edge and what should be examined first to improve the environmental efficiency of a target product.

The research results are used, for example, to distribute to in-house members of corporations for such purposes as working out corporate environmental management policies and for claiming, through public announcements, that company products are environmentally friendly.

LCA has in fact been in use since the 1970s. Then, the assessment method was mostly targeted at simple products such as containers for beverages. LCA is applicable to products ranging from office appliances to complex products such as railways and automobiles, as well as houses. The LCA is most successfully used when comparing different technologies providing the same benefit. That is, the most interesting results can be expected when major technology shifts happen in industry.

The standard for LCA has been internationally agreed [3]. However, the ISO standard does not give details for any product/technology groups but sets the minimum requirements for the execution of an LCA. The phases of the immensely data-intensive LCA tool are shown in Fig. 3.2.

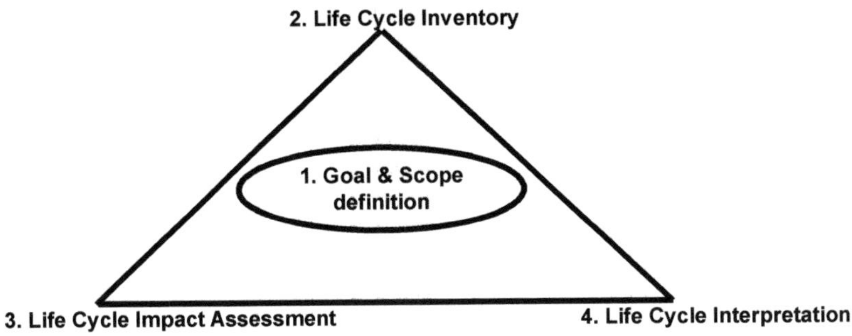

Fig. 3.2 The phases of LCA

LCA is assumed to consist of the four steps shown below.

1. *Setting of purpose and the scope of investigation*

The purpose of conducting the LCA is to be defined and the target scope of the investigation is to be decided. The scope of the investigation stated herein must include not only the range of assessment processes, but environment loading substances, environment impact categories and assessment models as well.

2. *Analysis of life cycle inventory (LCI)*

In the LCI the potential environmental loadings of all processes over the whole life cycle are calculated.

3. *Life cycle impact assessment (LCIA)*

In the LCIA the amount of potential environmental impact due to environmental loadings is assessed. LCIA commonly assesses the potential contribution to an environmental problem such as global warming.

In other cases, the LCIA integrates various types of environmental impact and expresses them as a single indicator. The way in which the assessment results are shown differs depending on the assessment method.

4. *Life cycle interpretation*

In the interpretation it is examined which process, substance and impact category is most important based on the LCI and LCIA. The reliability of data used is assessed and their precision by re-examination, if necessary. A conclusion must be drawn from the results.

The mathematical framework for the linear LCI/LCA model has been formulated and researched, e.g., by Heijungs [4], and Sakai [5]. Takahashi et al. [6] were inspired to set up the following equations:

$$g = BA^{-1}f \tag{3.1}$$

Where g is the vector of environmental loadings, B is the flow of environmental loadings, A is the flow of products and materials (e.g., electricity), f is the so called demand vector represented by a special unit process in which f_u is one of the outputs. E.g., in the solder/adhesive paper by Andrae et al. [7] the f_u was the amount of material needed to cover the PWB area in stencil printing.

The outcome of an LCI is a list (g_i) of emissions released and materials/energies consumed:

$$g = \begin{pmatrix} g_1 \\ \vdots \\ g_i \\ \vdots \\ g_n \end{pmatrix} \text{ for } i = 1,2,,,n \tag{3.2}$$

The LCIA is made up of selection, classification, and characterization [3]. Adding to the previous equations the LCIA part has the following formulations:

$$I_{i(j)} = g_i \times k_{i(j)} \; i = 1,2,,,n \tag{3.3}$$

$$I_j = \sum_{i=1}^{n} I_{i(j)} \; j = 1,2,,,.q \tag{3.4}$$

Where $k_{i(j)}$ are characterization factors, $I_{i(j)}$ impact indicators, and I_j the overall indicator of the environmental impact category. n and q are number of environmental loadings and number of environmental impact categories, respectively. For example, $GWP_{CO2(GWP100)}$ and $GWP_{CH4(GWP100)}$ could add up to GWP100 = $GWP_{CO2(GWP100)} + GWP_{CH4(GWP100)}$.

The last phase of LCA is the interpretation facilitated by normalization and occasionally weighting. For the normalization so-called reference values or normalization factors, w_j, are needed. Here follows two equations which define the "environmental burden" (EB):

$$I_{j(w)} = \frac{I_j}{w_j} \tag{3.5}$$

$$EB = \sum_{j=1}^{q} I_{j(w)} \tag{3.6}$$

For example, I_j/w_j could be the contribution of a certain environmental impact of a product targeted for assessment as the share of the UK total impact.

The unit of w_j could be "UK total impact per year" so therefore EB would be expressed as "years" or dimensionless. From this it is possible to identify to which impact category a target for assessment has made the largest contribution.

Heijungs treated unit usages within LCA [8]. Several models to arrive at subjective weighting factors for each $I_{j(w)}$ have been proposed.

One way was developed by Lin et al. where the weighting factors are obtained by the "distance-to-target" principle [9]. The distance to reach a certain target amount of, e.g., greenhouse gas emissions sets the weighting factor.

3.1.1 Allocation

Apart from selection of functional unit and system boundaries, the unavoidable allocation is up for most debate.

In LCA by allocation is meant the partitioning of the responsibility for environmental loading (the LCI) from the economic activities to a reference flow or a reference life cycle system [10].

Allocation should reflect the changes of product benefit over time. Product systems using recyclable material involve both multiple-input- and output-allocation [11]. As long as there is more than one useful input and more than one useful output from a system, allocation has to be performed.

Allocation can be avoided by expanding the system to include the life cycles of the inputs and output flows; however, that could lead to excessive data collection.

3.1.1.1 Multiple Input Allocation

An example of multiple input allocation is municipal waste treatment, e.g., at a landfill.

How could the environmental loadings caused by the landfill activities be divided among the incoming heterogeneous waste?

In Fig. 3.3 the incoming waste consists of five different types.

A situation could be that Waste 1 is the output from a household and Waste 2 from a building site, both sent to the same landfill site.

In an LCA study a practitioner hypothetically needs an LCI data model for the treatment of e.g., 1 ton of household waste sent for landfill, and likewise for building waste.

Then, the common parts of the landfill flows have to be allocated to each incoming waste to arrive at different models for household waste and building waste landfilling.

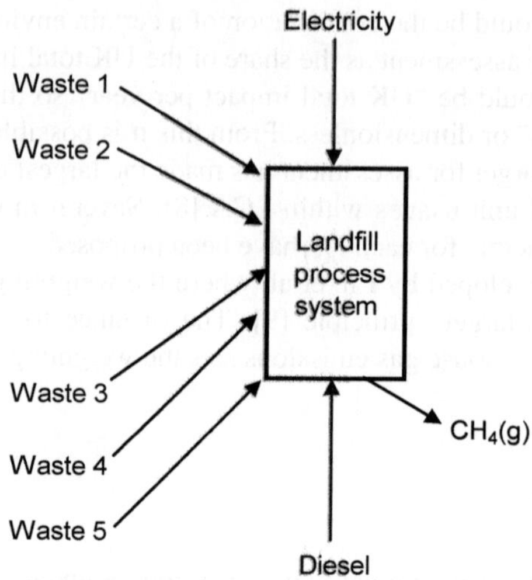

Fig. 3.3 A landfill as an example of input-based allocation in LCI

3.1.1.2 Multiple Output Allocation

Many factories produce more than one product and therefore the allocation of the environmental loadings becomes a constantly reoccurring problem.

The ISO standard does not specify which the preferable allocation method is, but realistically two main choices exist: causal (e.g., physical) and non-causal (e.g., economic).

Many variations of these alternatives exist and in Fig. 3.4 is an example of economic allocation for a Pb smelter.

The sensitivity of using different allocation methods should always be tested for the final LCI results.

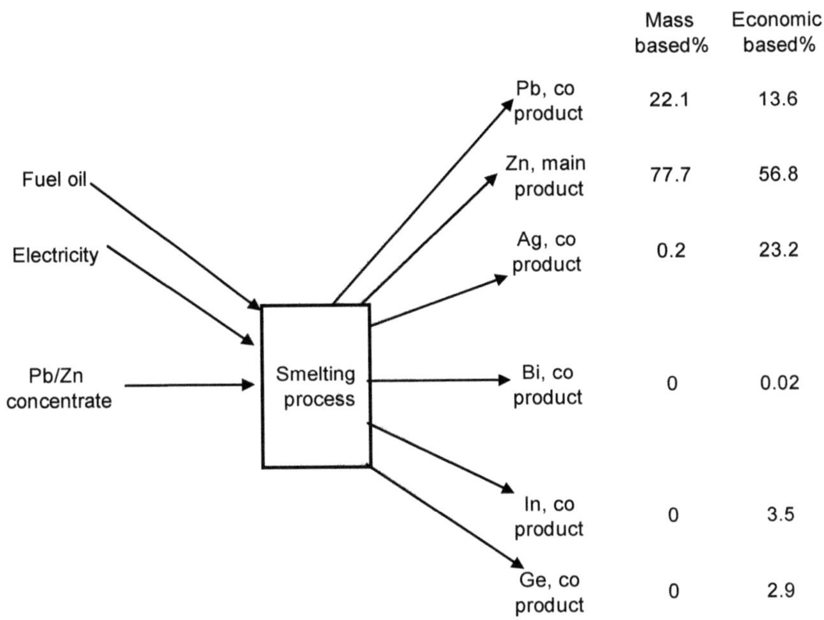

Fig. 3.4 A Pb/Zn smelter as an example of output-based allocation in LCI

3.2 Application of LCA

Although there are many problems associated with it [12,13], LCA has been successfully implemented in product eco-re-design, e.g., in the Hong Kong electronics industry for products exported to the European market [14].

Annually, several papers are published reporting the successful use of LCA. However, as LCA is a methodology that often involves proprietary process data, confidential reports are produced *within* companies. The results from these reports never reach the LCI databases used by other LCA practitioners.

3.3 Environmental Cost

A section on environmental cost is necessary as some of the weighting methods in LCA are based upon monetization of environmental damages. For example, CO_2 and SO_2 are generated by the oxidation of the carbon and sulfur content of petroleum and coal during the energy conversion. These substances cause environmental problems, such as global warming, acidification, and air pollution.

However, also processes such as natural pollution from volcanoes, and anaerobic processes in forests releasing gases such as methane, can circumstantially be characterized as environmental loadings.

These loadings can lead to environmental impact damages (midpoints) and finally environmental consequences in form of environmental costs and suffering (endpoints). Commonly LCA focuses on the anthropogenically induced environmental impacts. Moreover, several approaches attempting to quantify the endpoint environmental costs have been presented [15, 16]. The environmental costs can be characterized in at least five different ways as shown below in Table 3.1.

Table 3.1 Five different ways of describing environmental costs

Environmental cost	Scope	Comment
E1	The cost people are willing to pay to avoid, e.g., Pb being emitted into the environment	Least difficult to quantify
E2	The cost sustaining the environment	
E3	The economic costs actually caused by, e.g., air pollution	Health care, years of life lost, years lived with disability, material cost of corrosion etc.
E4	The cost of remediation	
E5	The cost that can be offset by not spending on the costliest control measures	

One (E1) is the *price people are willing to pay to avoid*, e.g., Pb being emitted into the environment.

Another environmental cost (E2) is the *price sustaining the environment*, e.g., the price sustaining the biodiversity in a forest compared to "avoiding" carbon emissions (leading to high risk of adding to the global heating and possible climate change) by producing biofuels from the trees in the same forest.

A third way is environmental cost 3 (E3) which is *the economic cost actually caused* by, e.g., air pollution in form of health care, years of life lost, years lived with disability, not mentioning the material cost of corrosion, etc. [17]. E1 and E3 were calculated in 2007 for coal usage in China by Chinese Greenpeace, World Wildlife Foundation, and the Chinese CSEP (The Energy Foundation) [18].

A fourth way is environmental cost 4 (E4) which is *the cost of remediation.* E4 is the cost of achieving a predetermined more desirable level of air, water, land and ecosystem quality as compared to the existing less desirable level [19]. E5 is the marginal emission control cost.

E2 is generally not quantified in society as many difficulties exist for allocating which actors should pay for which environmental damage, and moreover that data is difficult to obtain for land use etc. E4 is quantified for many nations as far as the cost for air, water, and solid wastes.

Many more data are likely needed to calculate E2–E4 as compared to E1. Nevertheless, it is agreed within the LCA research community that the uncertainty of the final E1–E4 scores is higher than for inventory scores. There has been an increasing awareness of formulating and quantifying the connection between the data input uncertainty and the output uncertainty. Several authors agree that the overall uncertainty increases rapidly for single weighting score comparisons, as opposed to comparing, for example emissions, of CO_2. E3 and E5 values were summarized in 2008 [20].

3.4 The Japanese LIME System for Life Cycle Impact Assessment

In 1998 as part of the national Japanese LCA project within the Japan Environmental Management Association for Industry (JEMAI), a new methodology for impact assessment was developed. A handbook in Japanese exists [21] and an English translation is under way, so the following introduction is all but full. The damage-oriented impact assessment method "Life cycle Impact assessment Method based on Endpoint modeling (LIME) expresses its single score according to E1.

LIME is a damage-calculation-type environmental impact assessment method based on the environmental conditions prevalent in Japan. Although LIME is limited to Japan, many of the details of its methodology of environmental impact assessment are applicable to other nations.

These are, e.g., results of assessment factors for LCIA of global phenomena such as the cost of global warming and ozone depletion.

Communicating environmental success information is getting more important for companies. LIME was developed as a method to assist the use of LCA, eco-efficiency, and environmental accounting efficiently.

The problem is that there has not been any LCA infrastructure available for such a use. To solve this sort of problem for Japan, the New Energy and Industrial Development Organization (NEDO) at the Ministry of Economy, Trade and Industry implemented a project to construct a database for LCA over five years from 1998 to 2003. This project was aimed at constructing the three kinds of systems listed below:

LCI data authorized by the 56 industry associations of Japan
Network system that allows the system user to obtain necessary data easily
Life cycle impact assessment method jointly developed by 30 environment scientists and academic experts

LIME is an impact assessment method that was developed through the implementation of the LCA project. Many companies have used *LIME* for LCA of their own products since the LCA project was carried out.

For example the electronics company Ricoh assessed its product lines to measure how the use of Pb-free solder would help reduce environmental impact. Their

LIME results indicated that a shift to Pb-free solder, preventing potential emission of toxic substances, could sharply reduce the environmental impact of the product life cycle.

For case studies it is often not enough to base decisions solely on CO_2 emissions. LIME consists of 11 impact categories and around 1,000 substances as assessment targets, giving the possibility of better understanding of the "total" environmental impact.

LIME in that sense supports the environmental accounting schemes introduced by many companies in Japan.

The attempt of monetize the total effects of environmental preservation would be attractive to users as they would be able to compare the social effects of environmental preservation with the direct costs.

Moreover, new product development costs vs. their environmental costs can be estimated using LIME.

LIME has already been used for assessment of full costs, environmental efficiency, and analysis of costs vs. environmental effects [22].

To make these analyses, it is necessary to calculate environmental loadings (or, in the case of LCA, to obtain inventory data). If this requirement is met, then assessment results can be obtained by corresponding factors and linear calculation.

LIME is divided into two steps. First, characterization and damage assessment is carried out, based on natural scientific principles. Second, integration is done through social science-based analysis.

A methodology for calculating the damage factors in damage assessment was developed.

A methodology exists for integration, the final step in LIME. In LIME, economic values for damage per end-point were estimated by conjoint analysis. Factors for obtaining dimensionless indicators were also calculated, taking into account the diverse purposes of LIME users.

As a summary a list of two types of factors, damage factors and integration factors, are shown in Tables 3.2 and 3.3.

Table 3.2 Selected LIME damage factors. CoV = Coeffcient of variation

Impact category (Damage category)	Impact on	Substance, Compound, Element, Material	Value, *t*	CoV%	Unit
Global warming (HH)	Air	Carbon dioxide	1.23×10^{-7}	65	DALY/kg
Global warming (HH)	Air	Dinitrogen monoxide	3.63×10^{-5}	68	DALY/kg
Global warming (HH)	Air	Methane	2.82×10^{-6}	68	DALY/kg
Global warming (HH)	Air	Sulfur hexafluoride	0.00272	64	DALY/kg
Global warming (SA)	Air	Carbon dioxide	0.548	71	JPY / kg
Global warming (SA)	Air	Dinitrogen monoxide	3.63×10^{-5}	58	JPY / kg
Global warming (SA)	Air	Methane	16.9	59	JPY / kg
Global warming (SA)	Air	Sulfur hexafluoride	0.00272	58	JPY / kg

Table 3.2 continued

Impact category (Damage category)	Impact on	Substance, Compound, Element, Material	Value, *t*	CoV%	Unit
Ozone layer depletion (HH)	Air	Ethane, 1-bromo-2-fluoro-, FC-151B1	9.84×10^{-5}	28	DALY/kg
Ozone layer depletion (HH)	Air	Ethane, 1-chloro-1,1-difluoro-, HCFC-142	8.24×10^{-5}	28	DALY/kg
Ozone layer depletion (HH)	Air	Ethane, 1,1-dichloro-1-fluoro-, HCFC-141b	1.2×10^{-4}	28	DALY/kg
Ozone layer depletion (HH)	Air	Ethane, 1,1,1-trichloro-, HCFC-140	9.13×10^{-5}	28	DALY/kg
Ozone layer depletion (SA)	Air	Ethane, 1-bromo-2-fluoro-, FC-151B1	6.61	53	JPY/kg
Ozone layer depletion (SA)	Air	Ethane, 1-chloro-1,1-difluoro-, HCFC-142	5.53	53	JPY/kg
Ozone layer depletion (SA)	Air	Ethane, 1,1-dichloro-1-fluoro-, HCFC-141b	8.03	53	JPY/kg
Ozone layer depletion (SA)	Air	Ethane, 1,1,1-trichloro-, HCFC-140	6.13	53	JPY/kg
Ozone layer depletion (PP)	Air	Ethane, 1-bromo-2-fluoro-, FC-151B1	21.2	95	kg/kg
Ozone layer depletion (PP)	Air	HCFC-142	17.7	95	kg/kg
Ozone layer depletion (PP)	Air	HCFC-141b	25.7	95	kg/kg
Ozone layer depletion (PP)	Air	HCFC-140	19.7	95	kg/kg
Photochemical oxidants (HH)	Air	NMVOC	6.99×10^{-6}	27	DALY/kg
Photochemical oxidants (SA)	Air	NMVOC	27.2	27	JPY/kg
Photochemical oxidants (PP)	Air	NMVOC	3.6	27	kg/kg
Acidification (SA)	Air	Ammonia	364	282	JPY/kg
Acidification (SA)	Air	Nitrogen oxides	43.4	339	JPY/kg
Acidification (SA)	Air	Sulfur oxides	60.8	364	JPY/kg
Acidification (PP)	Air	Ammonia	1.93	99	kg/kg
Acidification (PP)	Air	Nitrogen oxides	0.231	91	kg/kg
Acidification (PP)	Air	Sulfur oxides	0.323	126	kg/kg
Air pollution (HH)		Nitric oxide	1.46×10^{-5}	443	DALY/kg
Air pollution (HH)		Particulates	0.000253	247	DALY/kg
Air pollution (HH)		Particulates, < 10 um	0.000253	247	DALY/kg

Table 3.2 continued

Impact category (Damage category)	Impact on	Substance, Compound, Element, Material	Value, *t*	CoV%	Unit
Air pollution (HH)		Particulates, < 10 um (mobile)	0.00114	247	DALY/kg
Air pollution (HH)		Particulates, < 10 um (stationary)	0.000253	247	DALY/kg
Air pollution (HH)		Particulates, < 2.5 um	0.000416	235	DALY/kg
Air pollution (HH)		Particulates, > 10 um	0.000253	328	DALY/kg
Air pollution (HH)		Particulates, > 10 um (process)	0.000253	247	DALY/kg
Air pollution (HH)		Particulates, > 2.5 um, and < 10um	0.000253	268	DALY/kg
Air pollution (HH)		Sulfur oxides	0.000105	144	DALY/kg
Human Toxicity (HH)	Water	As	0.00148	700	DALY/kg
Human Toxicity (HH)	Soil	As	0.00265	700	DALY/kg
Human Toxicity (HH)	Air	As	0.000884	700	DALY/kg
Human Toxicity (HH)	Water	Cd	0.00694	700	DALY/kg
Human Toxicity (HH)	Soil	Cd	0.00909	700	DALY/kg
Human Toxicity (HH)	Air	Cd	0.0035	700	DALY/kg
Human Toxicity (HH)	Air	Dioxins, measured as 2,3,7,8-tetrachlorodibenzo-p-dioxin	132	700	DALY/kg
Human Toxicity (HH)	Soil	Dioxins, measured as 2,3,7,8-tetrachlorodibenzo-p-dioxin	1.57	700	DALY/kg
Human Toxicity (HH)	Water	Dioxins, measured as 2,3,7,8-tetrachlorodibenzo-p-dioxin	1,120	700	DALY/kg
Human Toxicity (HH)	Water	Pb	0.0474	700	DALY/kg
Human Toxicity (HH)	Soil	Pb	0.0359	700	DALY/kg
Human Toxicity (HH)	Air	Pb	0.0192	700	DALY/kg
Ecotoxicity (BD)	Water	As	3.07×10^{-8}	700	EINES/kg
Ecotoxicity (BD)	Soil	As	2.32×10^{-8}	700	EINES/kg
Ecotoxicity (BD)	Air	As	1.05×10^{-8}	700	EINES/kg
Ecotoxicity (BD)	Water	Cd	1.78×10^{-7}	700	EINES/kg
Ecotoxicity (BD)	Soil	Cd	1.34×10^{-7}	700	EINES/kg
Ecotoxicity (BD)	Air	Cd	6.07×10^{-8}	700	EINES/kg
Ecotoxicity (BD)	Water	Pb	1.37×10^{-8}	700	EINES/kg

Table 3.2 continued

Impact category (Damage category)	Impact on	Substance, Compound, Element, Material	Value, *t*	CoV%	Unit
Ecotoxicity (BD)	Soil	Pb	1.03×10^{-8}	700	EINES/kg
Ecotoxicity (BD)	Air	Pb	4.66×10^{-9}	700	EINES/kg
Euthropication (SA)	Water	Ammonium, ion	64.1	36	JPY/kg
Euthropication (SA)	Water	COD, Chemical Oxygen Demand	0.64	36	JPY/kg
Euthropication (SA)	Water	Nitrate	18.6	36	JPY/kg
Euthropication (SA)	Water	Nitrogen, total	82.5	36	JPY/kg
Euthropication (SA)	Water	Phosphate	318	36	JPY/kg
Euthropication (SA)	Water	Phosphorus, total	974	36	JPY/kg
Resource consumption (SA)	Natural resource	Al, in ground	0.0033	94	JPY/kg
Resource consumption (SA)	Natural resource	Sb, in ground	1.73	48	JPY/kg
Resource consumption (SA)	Natural resource	Bauxite, in ground	7.33×10^{-5}	97	JPY/kg
Resource consumption (SA)	Natural resource	Cr, in ground	0.0124	277	JPY/kg
Resource consumption (SA)	Natural resource	Coal, in ground	0.00934	14	JPY/kg
Resource consumption (SA)	Natural resource	Co, in ground	12.5	20	JPY/kg
Resource consumption (SA)	Natural resource	Cu, in ground	43	42	JPY/kg
Resource consumption (SA)	Natural resource	Gas, natural, in ground	0.691	75	JPY/kg
Resource consumption (SA)	Natural resource	Au, in ground	487,000	29	JPY/kg
Resource consumption (SA)	Natural resource	In, in ground	14,400	70	JPY/kg
Resource consumption (SA)	Natural resource	Fe, in ground	0.233	63	JPY/kg
Resource consumption (SA)	Natural resource	Pb, in ground	26.7	30	JPY/kg
Resource consumption (SA)	Natural resource	Mn, in ground	38.6	60	JPY/kg
Resource consumption (SA)	Natural resource	Hg, in ground	157	98	JPY/kg
Resource consumption (SA)	Natural resource	Mo, in ground	32.7	68	JPY/kg

Table 3.2 continued

Impact category (Damage category)	Impact on	Substance, Compound, Element, Material	Value, *t*	CoV%	Unit
Resource consumption (SA)	Natural resource	Ni, in ground	37.3	20	JPY/kg
Resource consumption (SA)	Natural resource	Nb, in ground	0.00273	38	JPY/kg
Resource consumption (SA)	Natural resource	Oil, crude, in ground	1.47	43	JPY/kg
Resource consumption (SA)	Natural resource	Ag, in ground	5,620	55	JPY/kg
Resource consumption (SA)	Natural resource	Sn, in ground	107	82	JPY/kg
Resource consumption (SA)	Natural resource	Ti, in ground	57.3	66	JPY/kg
Resource consumption (SA)	Natural resource	W, in ground	16.8	83	JPY/kg
Resource consumption (SA)	Natural resource	U, in ground	1,030	70	JPY/kg
Resource consumption (SA)	Natural resource	V, in ground	1.28×10^{-5}	218	JPY/kg
Resource consumption (SA)	Natural resource	Zn, in ground	42.5	42	JPY/kg
Resource consumption (PP)	Natural resource	Al, in ground	0.0726	121	kg/kg
Resource consumption (PP)	Natural resource	Sb, in ground	0.153	321	kg/kg
Resource consumption (PP)	Natural resource	Bauxite, in ground	0.00797	129	kg/kg
Resource consumption (PP)	Natural resource	Cr, in ground	0.0156	290	kg/kg
Resource consumption (PP)	Natural resource	Coal, hard, unspecified, in ground	0.0562	119	kg/kg
Resource consumption (PP)	Natural resource	Cu, in ground	0.378	151	kg/kg
Resource consumption (PP)	Natural resource	Au, in ground	940	406	kg/kg
Resource consumption (PP)	Natural resource	In, in ground	1.94	100	kg/kg
Resource consumption (PP)	Natural resource	Fe, in ground	0.00861	150	kg/kg
Resource consumption (PP)	Natural resource	Pb, in ground	0.0879	312	kg/kg

Table 3.2 continued

Impact category (Damage category)	Impact on	Substance, Compound, Element, Material	Value, *t*	CoV%	Unit
Resource consumption (PP)	Natural resource	Limestone, in ground	0.00258	60	kg/kg
Resource consumption (PP)	Natural resource	Mn, in ground	0.0168	319	kg/kg
Resource consumption (PP)	Natural resource	Hg, in ground	1.01	221	kg/kg
Resource consumption (PP)	Natural resource	Mo, in ground	4.11	256	kg/kg
Resource consumption (PP)	Natural resource	Ni, in ground	0.415	286	kg/kg
Resource consumption (PP)	Natural resource	Nb, in ground	0.951	325	kg/kg
Resource consumption (PP)	Natural resource	Sand and gravel, unspecified, in ground	0.00258	60	kg/kg
Resource consumption (PP)	Natural resource	Sand, unspecified, in ground	0.00258	60	kg/kg
Resource consumption (PP)	Natural resource	Ag, in ground	55.9	253	kg/kg
Resource consumption (PP)	Natural resource	Sn, in ground	12.5	298	kg/kg
Resource consumption (PP)	Natural resource	Ti, in ground	2.79	338	kg/kg
Resource consumption (PP)	Natural resource	W, in ground	0.697	269	kg/kg
Resource consumption (PP)	Natural resource	V, in ground	1.53	420	kg/kg
Resource consumption (PP)	Natural resource	Zn, in ground	0.107	225	kg/kg
Resource consumption (BD)	Natural resource	Al, in ground	5.15×10^{-14}	470	EINES/kg
Resource consumption (BD)	Natural resource	Sb, in ground	8.9×10^{-14}	1,017	EINES/kg
Resource consumption (BD)	Natural resource	Bauxite, in ground	5.65×10^{-15}	494	EINES/kg
Resource consumption (BD)	Natural resource	Cr, in ground	9.05×10^{-15}	823	EINES/kg
Resource consumption (BD)	Natural resource	Coal, hard, unspecified, in ground	2.73×10^{-14}	551	EINES/kg
Resource consumption (BD)	Natural resource	Cu, in ground	4.34×10^{-13}	511	EINES/kg

Table 3.2 continued

Impact category (Damage category)	Impact on	Substance, Compound, Element, Material	Value, *t*	CoV%	Unit
Resource consumption (BD)	Natural resource	Au, in ground	5.45×10^{-10}	1,176	EINES/kg
Resource consumption (BD)	Natural resource	In, in ground	1.64×10^{-12}	500	EINES/kg
Resource consumption (BD)	Natural resource	Fe, in ground	6.2×10^{-15}	496	EINES/kg
Resource consumption (BD)	Natural resource	Pb, in ground	5.1×10^{-14}	1,038	EINES/kg
Resource consumption (BD)	Natural resource	Limestone, in ground	1.97×10^{-15}	356	EINES/kg
Resource consumption (BD)	Natural resource	Mn, in ground	9.76×10^{-15}	1,289	EINES/kg
Resource consumption (BD)	Natural resource	Hg, in ground	5.87×10^{-13}	828	EINES/kg
Resource consumption (BD)	Natural resource	Mo, in ground	2.38×10^{-12}	790	EINES/kg
Resource consumption (BD)	Natural resource	Ni, in ground	2.41×10^{-13}	893	EINES/kg
Resource consumption (BD)	Natural resource	Nb, in ground	5.51×10^{-13}	829	EINES/kg
Resource consumption (BD)	Natural resource	Sand and clay, unspecified, in ground	1.97×10^{-15}	356	EINES/kg
Resource consumption (BD)	Natural resource	Sand, unspecified, in ground	1.97×10^{-15}	356	EINES/kg
Resource consumption (BD)	Natural resource	Ag, in ground	3.24×10^{-11}	762	EINES/kg
Resource consumption (BD)	Natural resource	Sn, in ground	7.26×10^{-12}	1,505	EINES/kg
Resource consumption (BD)	Natural resource	Ti, in ground	1.62×10^{-12}	1,028	EINES/kg
Resource consumption (BD)	Natural resource	W, in ground	4.04×10^{-13}	1,113	EINES/kg
Resource consumption (BD)	Natural resource	V, in ground	8.9×10^{-13}	1,056	EINES/kg
Resource consumption (BD)	Natural resource	Zn, in ground	6.22×10^{-14}	853	EINES/kg
Waste (PP)	Final waste	Carton waste	0.185	200	kg/kg
Waste (PP)	Final waste	Coal ash	0.0185	200	kg/kg
Waste (PP)	Final waste	Construction waste	0.0185	200	kg/kg
Waste (PP)	Final waste	Dust, unspecified	0.025	200	kg/kg
Waste (PP)	Final waste	Electronic waste	0.0449	200	kg/kg

Table 3.2 continued

Impact category (Damage category)	Impact on	Substance, Compound, Element, Material	Value, t	CoV%	Unit
Waste (PP)	Final waste	Fibre biomass waste	0.0786	200	kg/kg
Waste (PP)	Final waste	Fly ash	0.0185	200	kg/kg
Waste (PP)	Final waste	Glass waste	0.021	200	kg/kg
Waste (PP)	Final waste	Metal waste	0.0157	200	kg/kg
Waste (PP)	Final waste	Mineral waste	0.0163	200	kg/kg
Waste (PP)	Final waste	Plastic waste	0.105	200	kg/kg
Waste (PP)	Final waste	Printed circuit boards waste	0.0449	200	kg/kg
Waste (PP)	Final waste	Waste, from incinerator	0.0385	200	kg/kg
Waste (BD)	Final waste	Carton waste	6.7×10^{-13}	700	EINES/kg
Waste (BD)	Final waste	Coal ash	6.7×10^{-14}	700	EINES/kg
Waste (BD)	Final waste	Construction waste	6.7×10^{-14}	700	EINES/kg
Waste (BD)	Final waste	Dust, unspecified	9.05×10^{-14}	700	EINES/kg
Waste (BD)	Final waste	Electronic waste	1.63×10^{-13}	700	EINES/kg
Waste (BD)	Final waste	Fibre biomass waste	2.85×10^{-13}	700	EINES/kg
Waste (BD)	Final waste	Fly ash	6.7×10^{-14}	700	EINES/kg
Waste (BD)	Final waste	Glass waste	7.6×10^{-14}	700	EINES/kg
Waste (BD)	Final waste	Mineral waste	5.91×10^{-14}	700	EINES/kg
Waste (BD)	Final waste	Plastic waste	3.8×10^{-13}	700	EINES/kg
Waste (BD)	Final waste	Printed circuit boards waste	1.63×10^{-13}	700	EINES/kg
Waste (BD)	Final waste	Waste, from incinerator	1.4×10^{-13}	700	EINES/kg

Table 3.3 LIME weighting factors

Damage Category	Value (WF)	CoV%	Unit
Human health (HH)	9,700,000	48	JPY/DALY
Social assets (SA)	1	29	JPY/JPY
Primary productivity (PP)	20.2	43	JPY/kg
Bio–diversity (BD)	4.8×10^{12}	34	JPY/EINES

It is possible to conduct an analysis using LIME without full knowledge of the details. In LIME the normalization (Eq. 3.5) is done for the damage instead of the characterization meaning that the EB of Eq. 3.6 corresponds to HH, SA, PP or BD.

Below follows Eq. 3.7–3.10 which describe how the social cost (SC) is calculated. $t_{i(m)}$ is the damage factors, m is the damage category (Table 3.2), $D_{i(m)}$ the damage indicators, D_m is the overall indicator of the damage category, w_m is the normalization factor of the overall damage indicator, and $D_{m(wm)}$ is the normalized overall damage indicator.

For the normalization so-called reference values or normalization factors, w_m, are needed. r is the number of damage categories and WF the weighting factor of a damage category. Here follows four equations which define the SC:

$$D_{i(m)} = g_i \times t_{i(m)} \quad i = 1,2,\ldots,n \tag{3.7}$$

$$D_m = \sum_{i=1}^{n} D_{i(m)} \quad m = 1,2,\ldots,r \tag{3.8}$$

$$D_{m(w_m)} = \frac{D_m}{w_m} \tag{3.9}$$

$$SC = \sum_{m=1}^{r} WF_m \times D_{m(w_m)} \tag{3.10}$$

The main characteristics of LIME are summarized below.

3.4.1 The Framework for LIME

The framework of LIME is based on a top-down approach, contrary to the bottoms-up approach. Founded in the research field of environmental ethics, four so-called safeguard subjects were created: "human health," "social assets," "biovdiversity," and "primary production". Then, for each safeguard object, indicators representing the amount of damage resulting from the generation of environmental loadings were established.

The damage calculation of LIME can be followed with high transparency, and the major steps in LIME are described below in Fig. 3.5. For "human health", disability-adjusted life years (DALY) is used internationally as an indicator to measure the loss of life expectancy by the World Health Organization (WHO), and was therefore chosen as ideal for the LIME system.

For "social assets", economic indicators that can comprehensively describe the impacts on a variety of constituent elements such as agricultural resources, marine resources, and mineral resources (Japanese currency, Yen), were integrated. "Bio-diversity" is in LIME expressed in "Expected increase in number of extinct species" (EINES), obtained from the summation of the extinction probability estimated for each species of organisms. "Primary production" is expressed as "Lost amount of net primary productivity" (NPP) widely used as an indicator to represent the wealth of an ecosystem.

In LCA methodology it is also possible to define the cause effect chain in six entities as inventory, midpoint environmental category, midpoint environmental indicator, endpoint environmental category, endpoint environmental indicator, single score. For example, as shown in Fig. 3.6, CO_2, GWP100, infrared forcing toward Earth, infectious diseases, DALY, Japanese yen. Additionally in LIME, 11 environmental problems closely related to the four safeguard subjects were defined as impact categories: i.e., global warming, ozone layer depletion, acidification, urban air pollution, hazardous chemical substances, ecotoxicity, photochemical oxidants, eutrophication, land use, resource consumption, and waste. The damage generated by these impact categories is analyzed and evaluated for each damage aspect before being aggregated in a common endpoint. In LIME, the damage aspect is called the category endpoint.

The scope of assessment targets in LIME is defined through so called category endpoints. LIME allows users to get LCIA results in three steps: characterization, damage assessment, and integration. These three steps are used for efficient life cycle interpretation of the results.

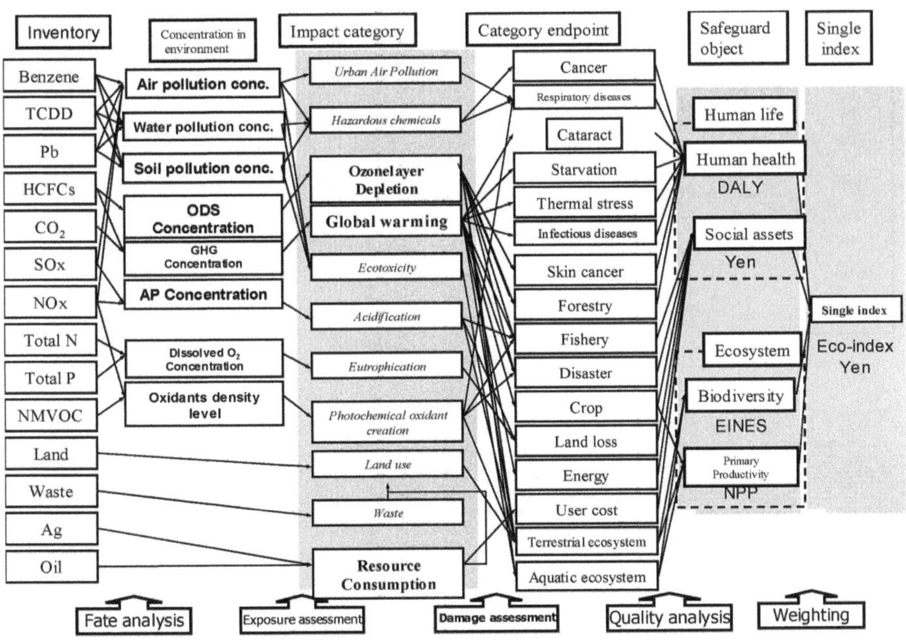

Fig. 3.5 The principal framework of LIME. ODS = ozone depleting substances, GHG = greenhouse gas, AP = acidifying pollutants

LIME therefore makes it possible to conduct an LCA that complies with the LCIA guidelines of the standard ISO 14042.

3.4.2 Characterization

ISO defined characterization as a requirement for LCA. Since the first half of the 1990s the development of characterization factors has been conducted internationally and has a longer history than the other areas in LCIA. In LIME the existing research achievements were first investigated and then the characterization factors that were most relevant for LCA users in Japan were selected.

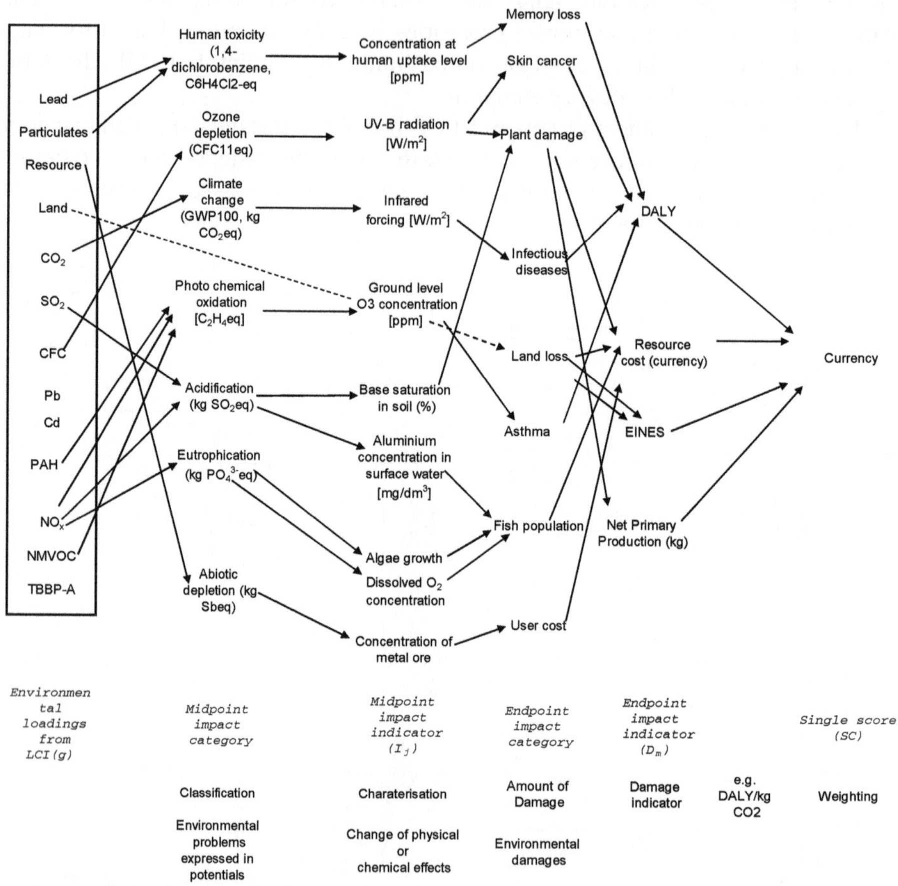

Fig. 3.6 Some cause-effects links in LCIA

Regarding global impact categories such as global warming, ozone layer depletion, and resource consumption, various factors with a high versatility had already been published by international institutes. Thus, after clarifying the differences among those factors, the developers of LIME selected recommended factors.

Since the regional and local environmental impacts are significantly affected by the environmental conditions of the sites where environmental loadings are generated (climate and geography, etc.), characterization factors based on Japanese environmental data were developed, followed by the setting of recommended factors compared to existing research.

3.4.3 Damage Assessment

The construction of damage functions that quantitatively relate inventories to category endpoints is crucial to the development of damage factors that establish a relationship between inventories and safeguard subjects. In LCIA research, the study of construction of damage functions is still a relatively new field. On the other hand, the damage functions are fundamental to LIME and thus research and development in this field was emphasized. The LIME model is based on connections between, e.g., emissions and their endpoints, and by weighting impact categories or damage categories the overall single index can be expressed in yen. One example is the emissions of NO_x which lead to acidification of freshwater systems and eutrophication of aquatic ecosystems, and which can also reach dangerously toxic levels for aquatic animals. The connection between NO_x emissions and adverse effects on human health and economy has been identified by research [23] by which the LIME method is constantly updated.

Some impact categories have many endpoints and LIME cannot cover them all. Prior to the development of damage functions, a screening survey to identify important category endpoints was done.

One of the most important contributions of LIME to LCIA research is that it established damage assessment as an additional step.

First, by making use of natural scientific principles, comparisons among various environmental impacts became possible, while minimizing the introduction of value judgments. For example, comparisons between CO_2 and chlorofluorocarbons (CFCs) have been conducted in the past by using weighting based on a subjective value judgment among environmental problems such as increased global warming and ozone layer depletion. The damage assessments in terms of the increments of thermal stress and malaria caused by increased temperature or the increment in number of cataracts and skin cancers caused by ultraviolet exposure can be compared and integrated using the loss-of-life expectancy. These indicators have been previously studied in actuarial statistics and ecology, and thus the effective utilization of these research achievements enabled logical, consistent, and reliable results.

Second, by setting endpoints as the assessment targets, the number of items in the results was reduced. This made it possible to avoid as much as possible the occurrence of trade-offs between the assessment results. In addition, the minimization of the number of items that are the targets of weighting is an important requirement for the development of integration factors with high social consensus. In LIME, a total of five items – four environmental attributes and one currency attribute – are applied to conjoint analysis, whereas the results of damage assessment are used for information regarding environmental attributes. Since it has been pointed out that humans cannot make rational judgments on more than six items at a time, damage assessment can be said to be a critical step in the integration of social consensus.

Third, assessment results positioned amidst the characterization and integration become available. Damage assessment that as far as possible avoids value judgments and minimizes the number of items has advantages in its reliability and ease of decision-making, and is expected to be used to complement the assessment results of characterization and integration.

Fourth, damage assessment reflects the previous achievements of fundamental research in environmental science.

Knowledge such as dose-response functions in epidemiology and toxicology, fieldwork in agricultural science and ecology, and the computation of extinction probability in mathematical biology constitute highly valuable information for LCIA.

Nevertheless, key knowledge has not been sufficiently utilized in previous LCIA methods.

The adoption of damage assessment has enabled the above-mentioned research achievements to gain an understanding of the relationship between changes in the environment and the corresponding response of endpoints. As a result, the LCA interface is fundamental research of the link between environmental science and product assessment.

This is an important achievement in reflecting the most recent knowledge in each field where continuous further development is expected in LCIA in the future.

Lastly, the results of damage assessment are unequivocal compared to characterization. Taking the characterization of acidification as an example, the ratio between the deposition of acid per unit amount of emission and that of SO_2 represents a result.

On the other hand, in damage assessment, the results are expressed as a decrement of plant production (kg) or amount of loss of marine resources (Japanese yen), making the meaning of the results easy to comprehend. Consequently, it is useful for verifying results and is additionally valuable as a communication tool.

It was crucial for the development of damage functions how scientific knowledge in fields like ecology, toxicology, epidemiology, and meteorology was collected. Moreover, the ease which the developers of LIME could feed the data into LCIA methods was important. An expert panel consisting of specialists in each field was organized for these purposes.

Through discussions by the panel, cooperation with experts in environment fields other than LCIA was stimulated and the latest knowledge from each field was successfully incorporated into the damage functions as components.

3.4.4 Integration

Integration was conducted using conjoint analysis, regarded as one of the latest assessment methods within environmental economics. The theory of this method is based on environmental economics and inferential statistics. Although there have been numerous records of application of conjoint analysis as a method of measuring partial effects in ecosystems, LIME was he first application within LCIA.

In conjoint analysis, the weights between the compared items are calculated by analyzing the results of a questionnaire survey targeting general consumers. For the questionnaires used in surveys, indicators that show the names of compared target items and quantitatively represent the condition of each item are crucial. In LIME, safeguard subjects and normalization values (potential damage that can be generated as a result of annual economic activity in Japan) were used.

Questionnaire surveys included mail surveys, interviews, and Internet surveys. In LIME, the respondents' understanding of the content of a questionnaire was prioritized, and thus interview surveys were employed. Discussions and preparatory tests were repeated to prepare a questionnaire that was comprehensible for respondents ahead of the implementation of the main investigation.

In the survey, which targeted 400 males and females aged 20–50, appropriate measures were taken not to bias age composition, gender, or annual income. The statistical superiority and the explanatory power of the logic model used for the analysis of the obtained result were both adequate, and integration factors applicable to general-purpose uses were obtained.

In LIME, the following three lists of integration factors were published:

1. external costs obtained by conjoint analysis;
2. nondimensional integration factors obtained by conjoint analysis; and
3. nondimensional integration factors obtained by analytic hierarchy process.

Users can select a type suited to their purposes from these lists and also can use it for sensitivity analysis to verify the influence of different characterization factors.

The use of the integration of environmental impacts (the amount of damage) has recently been attracting attention, not only in LCIA but also in environmental accounting and corporate assessment through environmental reports.

The framework of LIME would be applicable to any other nation than Japan. The same is true for the Swedish Environmental Priority Strategies in Product Development (EPS) method intended for global usage [24]. Eco-Indicator '99 is a well-known weighting method for Europe but its single score, points, is not directly interpretable in monetary units [25].

Featuring the LIME impact assessment method, sensitivity and uncertainty analyses will be presented in Sect. 6.4 for total system LCAs, where the impact assessment uncertainty is included. The LIME results could be compared to EPS, a weighting method which also includes uncertainty information. For the endpoint method LIME so far no study which includes "total" uncertainty has been presented.

LIME can be used to reflect both effects related to energy use and possible toxicity impacts mainly from unconventional metal recovery and uncontrolled disposal of electronic products.

3.5 Previous Research on LCA of Interconnection Materials

Since the middle of the 1990s there has been an increasing research interest in the development of environmental assessment methodologies and models to be used for electronic products including interconnection material technologies (IM). Low [26] concluded for telecom systems that large benefits are achieved even with simple approaches, which make the use of LCA methods justified.

Several case studies have been published, where IM products and processes have been environmentally assessed, both through attributional LCA (ALCA) and by means of other environmental assessment models. The reason for this review was the need to establish the scientific knowledge situation regarding the environmental effects and consequences of the shift to Pb-free IM.

3.5.1 LCA-based Approaches

No overall approaches based on LCA were found in the peer-reviewed literature aiming at systemizing, developing and simplifying the LCA work solely for IM. However, Vignes proposed a generic and simplified screening LCA methodology which has yet to be used for IM [27].

3.5.1.1 LCAs of Solder Interconnection

In 2004 Reuter et al. wrote two innovative papers about the dynamic transition from Pb-based solders to Pb-free [28, 29]. The approach published in two papers is of top-down policy character, based on material flow analysis. It was proposed that a dynamic modeling, as a life cycle inventory of the total system, is required to assess the environmental aspects of Pb-free solders. The question was raised, whether a ban of Pb would be an environmental benefit, as the manufacturing of solders could be considered a global open-loop recycling and production system.

The governmental decision of introducing Pb-free solders will bring the system out of steady state. Globally, production of Pb, Cu and Sn will not be affected significantly but locally, changes could be important. Eco-Indicator '99 scores were higher for Sn–Ag–Cu than for Sn–Pb. The LCIA findings indicated that preventing human toxicity will lead to resource damage. Important concluding remarks were (1) there is limited production infrastructure for co-products in Pb ore processing, (2) the banning of Pb should be restricted to products which are not likely to be recycled, to decrease Pb in landfills.

Moreover, the authors estimated global transport distances from the average distances between countries mining metal ores and countries producing crude metals, and the import and export data of unwrought refined metal and alloys, and of metal scrap. Another prediction was that the Pb demand will decline and the Sn demand increase. Dependent on substitute solders, the demand for bismuth, copper, zinc, silver, and antimony rises. As an estimate of the impact of the introduction of Pb-free solder, the difference between the base scenario and the Pb-free scenarios is used. The results indicate that that the Eco-Indicator '99 LCA scores for a number of Pb-free solders are higher than for the conventional solder. This is partly due to the higher environmental impact of Sn resource depletion compared to Pb resource depletion: all Pb-free solders have higher Sn concentrations than Pb based. The Ag-containing solders have *lower* impacts, due to the interconnections between Ag and Au production: Ag produces raw materials for Au production.

The authors argued that an increase in Ag production will lead to an increase in intermediate (152.0% for SnAgCu) and thus to a reduction of the amount of Au ore that is required, but also to a reduction of Au intermediate produced (92.9%). Because of the high value of Au ore for resource depletion, this leads to lower resource depletion values. LCA is a subjective process as it is often the perception of the relative importance of the different environmental problems that in the end determines the environmental performance, or ranking, of alternatives. The Eco-Indicator '99 scores were dominated by resource depletion. Sometimes, however, resource depletion is considered more of an economical than environmental problem. Moreover, the uncertainty in the "surplus energy method" used in the Eco-indicator '99 may lead to overestimation of the mineral depletion values, in particular, for Au and Ni. If resource depletion was not considered, then all Pb-free solders reduced the environmental impact by the same order of magnitude; the Ag containing solders scored the worst and the Sn–Cu solders scored the best. If the weighting of resource depletion was reduced to 5%, all Pb-free solders score better than conventional solders. The Ag-containing alloys score better than the other Pb-free alternatives. The impact of the three alternatives is on the same order of magnitude, the Sn–Zn–Bi alloy having the lowest impact. This shows that in order to evaluate the environmental impacts of a change, an understanding of both the industrial and natural resource cycles is critical.
According to Reuter, if Pb is substituted by Bi or Ag, the effect of the transition on the overall operation will be significant.

The simulation results showed that a substitution by Sn–Ag–Cu would increase Ag ore consumption by more than 9%, as reported in the literature, and that recycling must increase significantly to counter this effect. For Bi, the situation was even more pronounced, as the metal has no dedicated ores, and Pb production processes are one of the main sources of raw materials for Bi. As the Pb intermediates (Kroll–Betterton dross and anode slimes) are prime feedstock for Bi producers, their decreased availability forces Bi producers to look for, and switch to, alternate feedstock. In the base year 2000, only a small portion of the Bi production came from other intermediates in proportion to the Pb intermediates, and it is doubtful whether a sufficient amount of these intermediates will be available to enable a full transition to Sn–Zn–Bi solders. Thus, intermediates from the production of other metals must be found, for example, from Sn production. This example by Reuter et al. showed a paradox: Bi is to replace Pb in solder, but Pb is its source. In the model, this paradox was solved by assuming that new resources for Bi can be found. If Bi is produced from intermediates, the effect on resource depletion will be minimal.

In 2004, the Intelligent Manufacturing Systems (IMS) project EFSOT "Next Generation Environment-Friendly Soldering Technology" was partly performed in Japan. Using the LIME methodology, Itsubo et al. identified which issues were significant for solders 63Sn–37Pb, 96.3Sn–3Ag–0.7Cu and 99.3Sn–0.7Cu [30]. Setting the system boundaries from cradle-to-gate the *human health damage* was found potentially to be reduced by 40–50% after the shift to lead-free. The system boundaries were expanded to also include the disposal stage. Then the 63Sn–37Pb was found to have many negative consequences, especially when illegally dumped. However for other end-of-life options the trade-off between the social assets (Ag and Sn consumption) and the human health damage (Pb emissions) became more apparent. An important task was identified to be the design of a collection and recycling system for Pb-free solders.

In 2005 USEPA performed one of the most comprehensive LCAs of solders ever [31]. The functional unit was 1,000 cm^3 solder applied to a printed wiring board and the system boundary was from cradle to grave. The authors compared the relative impacts of Sn–Pb solder and Pb-free alternatives and the selected solder pastes were 63Sn–37Pb (Sn–Pb) 95.5Sn–3.9Ag–0.6Cu (SAC), 42Sn–57Bi–1Ag (BSA), and 96Sn–2.5Ag–1Bi–0.5Cu as reflow application solders. Bar solders examined were 63Sn–37Pb, 99.2Sn–0.8Cu, and 95.5Sn–3.9Ag–0.6Cu. Primary data were collected from manufacturers of solder paste, powder, flux, and reflow manufacturing and end-of-life. The solder pastes were evaluated for 20 different impact categories, e.g., global warming, air particulate impacts, occupational health, public human health, and aquatic ecotoxicity. The results suggested that BSA and 99.2Sn–0.8Cu are the most preferable alternatives for reflow and wave solder, respectively.

In 2005 *EFSOT* in Europe published their major work as a report [32]. The functional unit was 0.11 cm^3 of Sn–Pb solder and finish on 1 cm^2 of printed wiring boards for wave and for reflow soldering. *0.11 cm^3 is the volume of 1 g 63Sn–37Pb solder alloy, not 63Sn–37Pb solder paste.* The system boundary was from cradle to grave. Unfortunately their results are only given as single scores of the single score Eco-Indicator '99 weighting methodology, preventing the calculation of CO_2e per mass or volume of solder.

In 2006 Andrae and Ekvall attempted consequential LCA of the shift from 62Sn–36Pb–2Ag solder paste to 95.5Sn–3.8Ag–0.7Cu solder paste [33]. The functional unit was 0.53 cm^3 solder paste applied to a printed wiring board and the system boundary was from cradle-to-grave.

In 2008 the LCI database ecoinvent published cradle-to-gate data about solders [34], e.g., "Solder, bar, Sn63Pb37, for electronics industry, at plant/GLO U".

3.5.1.2 LCAs of Polymer Interconnection

In 1996 Segerberg and Hedemalm [35] reported an LCA comparison between Sn–Pb solder and Ag-epoxy adhesive. It is not specified exactly which material compositions are compared, but it can be assumed that it is 63Sn–37Pb and 80Ag–ICA. In contrast to Conway (Sect 3.5.3.2), it is argued that Ag from adhesives would be possible to recycle. The functional unit is the attachment of 1,888 joints of quad flat pack-circuits. It was assumed that 0.5 mg solder was needed per joint, but only 0.08 mg adhesive. All in all this paper can not be evaluated due to lack of details, however, the results of the very simple LCA are mostly dependent on the allocation method used in metal production, as well as assumptions about the waste treatment method.

Peer-reviewed environmental assessment studies featuring ECAs include Liu et al. who in 1998 performed an environmental evaluation of an electronic package where tremendous benefits would be realized if using flip-chip and an ACA film instead of a quad flat packagechip microcircuit and LMP Pb-based solder [36]. However, the end-of-life phase was omitted from that study.

In 2000 the Danish Toxicology Center performed a comparative LCA of electrically conductive adhesive and traditional Sn–Pb solder [37]. The results are presented as the difference between the adhesive and the Sn–Pb solder. On top of this the Danish weighting method Environmental Design of Industrial Products (EDIP) was used, showing no inventory or CO_2e data but only weighted result. EDIP uses the "unit" person-equivalent-target.

In 2007 and year 2008 Andrae published research comparing high-temperature materials [38, 39].

The CO_2 equivalent (CO_2e) results for the above case studies of solders and adhesives are given in Table 3.4.

Table 3.4 Results of selected CO_2e cradle-to-gate assessments and LCAs of solder and adhesive interconnection

Material, system boundary, reference	kg CO_2e /kg
Solder, reflow paste, 63Sn–37Pb, cradle-to-grave, [31]	174 (817 kg per 1,000 cm^3) density 4.7 g/cm^3
Solder, reflow paste, 95.5Sn–3.9Ag–0.6Cu, cradle-to-grave [31]	198 (873 per 1,000 cm^3) 4.4 g/cm^3
Solder, reflow paste, 42Sn–57Bi–1Ag, cradle-to-grave [31]	180 (631 per 1,000 cm^3) 3.5 g/cm^3
Solder, reflow paste, 62Sn–36Pb–2Ag, cradle-to-gate [33]	22
Solder, reflow paste, 95.5Sn–3.8Ag–0.7Cu, cradle-to-gate [33]	30
Solder, bar, Sn63Pb37, for electronics industry, at plant/GLO U, cradle-to-gate [34]	11.4

Table 3.4 continued

Material, system boundary, reference	kg CO_2e /kg
Solder, bar, Sn95.5Ag3.9Cu0.6, for electronics industry, at plant/GLO U, cradle-to-gate [34]	20.5
Solder, paste, Sn63Pb37, for electronics industry, at plant/GLO U, cradle-to-gate [34]	17.1
Solder, paste, Sn95.5Ag3.9Cu0.6, for electronics industry, at plant/GLO U, cradle-to-gate [34]	26.2
42Sn-58Bi-ICA, cradle-to-grave [39]	21.9
80Ag-ICA, cradle-to-grave [39]	207
10Sn-90Pb solder paste, cradle-to-grave [39]	38

Nakamura et al presented a hybrid input–output (IO) model for Japan, based on a waste IO model, which displays the mutual interdependence of metal production processes for many metals used in IM. The model was tested for a hypothetical case where in one of the scenarios 95.5Sn–3.9Ag–0.6Cu Pb-free solder is introduced in Japan replacing 63Sn–37Pb. One of the results are that the Ag demand would increase by more than 50% and Sn by more than 20%. Nakamura et al. also discussed whether recycling will be enough to meet Japan's Ag demand [40]. Nakamura's research considers the interdependence between the flows of products and wastes within a nation. The models are used in emerging research fields about optimizing recycling systems concerning several nations [41].

3.5.2 *Other Types of Environmental Assessment Approaches*

Several different methods and tools which can be used for general environmental assessment of IM were found in the literature. They resemble LCA approaches, but are not equivalent to the attributional LCA approach, as they are more simplified.

In 2003, Fukuda et al. predicted, based on the Japan Electronics and Information Technology Industries Association (JEITA) 2002 roadmap, that by the end of 2005 the Japanese transition to Pb-free products would be complete [42]. This is due to laws promoting environmentally friendly material usage and compliance with end-of-life recycling requirements. Moreover, Japanese companies have discovered that "green" products lead to increased market shares, e.g., once marketed as being green, a Matsushita mini disc player rose from 4.7 to 15%.

Japanese interests in Pb-free solders include, e.g., Sn–Ag–Cu alloys and the USA based National Electronics Manufacturing Initiative (NEMI) recommended 95.5Sn–3.9Ag–0.7Cu.

All in all, Fukuda et al. argued that Japanese electronics companies, through voluntary projects reducing and eliminating the toxic substances early in the product development process, were aiming to take the lead in electronics packaging with Pb-free initiatives.

The German Fraunhofer Institute suggested the Environmental Engineering Toolbox, as obstacles were found to the application of full-scale LCA of IM [43]. An important part of the Environmental Engineering Toolbox is a method for assessing maximum toxicity, the toxic potential indicator (TPI). It has been applied to discarded mobile phones in China [44].

The TPI is also used in electronics manufacturing [45] and it captures the environmental aspect of toxicity, which is a common concern of legislation and companies.

Several references can be found, especially to conference proceedings, where the TPIs for different solders have been calculated.

Lichtenvort et al. presented a method which simultaneously covers environmental, economic and legal aspects for electronic products affected by the WEEE and RoHS legislations. The IMs are one of the foci of RoHS. The method is called "grEEEn" and is intended to support environmental experts and design teams [46].

Park et al. proposed an environmental assessment tool for use in eco-design of consumer electronics of which IMs is one part [47].

Zhou and Schoenung suggested an integrated methodology for evaluation of technology shifts such as the Pb-free shift in the solder business [48].

The methodology is capable of making trade-offs between technological, economic and environmental factors and facilitates the quantification of the uncertainty induced by the value judgement.

The above approaches are intended for "design-for-environment" of microelectronics. The models were designed to mitigate the supposed poor compatibility of full LCA in product design.

3.5.3 Case Studies of Interconnection Materials Using Other Methods than LCA

The literature contains a few case studies where other approaches than LCA were applied to IM to explore their environmental properties.

3.5.3.1 Solders

In 2000, Griese et al. [50] summarized the Fraunhofer IZM studies concerning the implementation of Pb-free soldering. A TPI evaluation indicated that 63Sn–37Pb had a value of 8.5 TPI/mg and 96.5Sn–3.5Ag had 2.5, whereas the TPI-value for Bi, to be used for 42Sn–58Bi, could not be calculated due to missing information for Bi. This means that *Bi could be underestimated in TPI evaluations*.

On the other hand, for Ag the TPI value can be considered an overestimation as the legal threshold values are very strict to prevent the disease *argyria*. Regarding the impact of mining and metal production, an evaluation using the Eco-Indicator '95, the 96.5Sn–3.5Ag had 5% and 42Sn–58Bi had 62% of the value of 63Sn–37Pb. Bi was based solely on it being a by-product of Pb production, causing a relatively bad rating. As for the "disposal behavior" of soldered test PBAs, the toxicity characterization leaching procedure (TCLP), method 1311 confirmed that Pb was the highest priority as a pollutant, while Sn and Ag leached to a very low degree. The authors conclude that there were no major technical problems with Pb-free interconnection, and that full LCAs would uncover the environmental behavior of the new materials. These authors were under the impression that Pb had to be banned from interconnection systems around the world to prevent uncontrolled emissions of Pb, and that it was clear that Pb-free alloys would decrease the potential environmental impact of electronics.

In 2001, Hamano et al. [51] performed a "sustainability" evaluation comparing 95.8Sn–3.5Ag–0.7Cu to 63Sn–37Pb regarding energy and metal usage. The energy required for metal (Sn, Ag, and Cu) production was predicted to be only 15% of the reflow energy of 95.8Sn–3.5Ag–0.7Cu. The authors argue that although Pb-free soldering will imply increased reflow energy usage, it will still only occupy 1% of the energy usage of the electronics industry. The resource scarcity issue was assumed to be a more severe sustainability problem than the energy issue, however not a dramatic one, as Ag was assumed to be recycled from printed board assemblies.

In 2001 Turbini et al. endorsed a top-down analysis of the industrial ecological implications of the Pb-free shift [52]. An "environmentally responsible product assessment matrix" (Table 3.5) was set up and 63Sn–37Pb solder systematically compared to Pb-free 95.5Sn–3.9Ag–0.6Cu. The authors presented results showing that in 1998, Pb was not a concern as a pollutant from US landfills, and in turn PBA electronics was a small share of the municipal solid waste.

Mixed results were discussed regarding leachability of Ag and Pb from solid metal spheres, solder wire, and PBAs, so no definite conclusion could be drawn regarding the priority of Ag or Pb as possible pollutant from landfills. *In Chap. 7 of the present book, different leaching tests of PBAs will be discussed.* One important recommendation was that conventional Pb-based electronics recycling should be increased instead of introducing Pb-free solder. The main reason is to recycle Cu in order to decrease environmentally damaging primary Cu production and Cu ore mining. In the sense that the WEEE recycling should be increased, the authors agreed with the intention of the European WEEE legislation.

Table 3.5 Environmentally responsible product assessment matrix for interconnection materials

	Environmental parameters (change) →				
	Materials choice	Energy use	Emissions to		
Life cycle stage affected ↓			soil	water	air
Ore mining/metal production	Interwoven global metal production, Pb mining will not change substantially, increased Sn mining	Increase	Increase	Increase	Increase
Production	Different flux	Increase	–	–	–
Distribution	–	Small decrease due to lower mass	–	–	–
Use of IM: reflow/wave soldering	Weight applied will decrease	Increase	–	–	–
Disposal of e-waste	–	–	Small changes	Small changes	Small changes

In 2002 Graedel made the case that Bi and Ge supplies will quickly be threatened if 60Sn–40Pb solder would be universally replaced by the quinternary 93Sn–4Bi–2Ag–0.5Cu–0.1Ge alloy. A logic sequence was presented where eight questions are asked about the material a designer wants to substitute, and the replacement material; e.g., one question is about the possible supply problems of the replacement material [53]. Moreover, Graedel also argued that under the current economic growth concept, material recycling will only slow down the "depletion", not reverse the material consumption increase.

In 2003 Ku et al. presented a metrics evaluation of the most common individual metals (Pb, Ag, Cu, Sn, Sb, Bi, In) which make up solders. The metrics had three parts; toxicity, availability, and "environmental impact" of extraction. The metal with the lowest score (e.g., Pb most toxic, place 1, score 1; indium (In) least availability place 1, score 1; Ag most "environmental impact" at extraction, place 1, score 1) was the least preferable. The results show that Cu, Sn, and Pb were the most preferable and Ag and In the least. Bi was recognized as a coming resource supply problem [54].

In 2003 Mihaescu and Carstea argued that the main obstacles hindering the shift to Pb-free are the high production costs and the limited global resources of some materials [55]. The authors observed that no single replacement for the 63Sn–37Pb alloy existed, which would not cause problems in the automatic soldering equipment's logistics.

In 2007, Deubzer summarized his findings in a Ph.D. thesis [56]. The approach was a top-down using as far as possible global total statistics. One of the main conclusions is that introduction of Pb-free solders and finishes substantially and effectively will reduce the toxicity and the possible disposal risks related to IM in WEEE. At the same time, compared to conventional Sn–Pb soldering materials, the Pb-free materials double the manufacturing cost for EEE manufacturers, increase the energy usage and the loss of valuable resources. Pb-free IM are less efficient in resource and energy use.

In 2008 Zhou and Schoenung applied their integrated methodology [49] to the results from USEPA solder LCA [31]. That LCA reported results from 16 environmental impact categories which can be weighted into a single score. These aggregated total environmental impact scores indicated that Sn–Cu was the most preferred Pb-free bar solder, while Bi–Sn–Ag and 96Sn–2.5Ag–1Bi–0.5Cu are both preferred as paste solders. Over the whole life cycle, SAC had environmental impact comparable to that of 63Sn–37Pb.

3.5.3.2 Conductive Adhesives

In 1993 Conway [57] reported about a qualitative life cycle comparison between solders and adhesives for SMT interconnection. Conway speculated that more Pb will be produced to extract sufficient quantities of Bi for adhesives and solders. The environmental impact from high-volume production of adhesives was unclear as these products have been made only in relatively small quantities. Solders and adhesives both used similar processes regarding manufacturing of particles/flakes and for mixing. Flakes were made by rolling particles flat. Metal recovery from adhesives may be problematic and maintenance/repair difficult. Heat or softeners could be needed to degrade or soften the adhesive. The major result was that not enough knowledge existed about candidate materials replacing Sn–Pb solders.

In 1999 Mueller et al. [58] set up material and energy balances and compared all differences between 63Sn–37Pb solder paste and Ag-filled epoxy for SMT. The reference measure was the daily production of a PWB mounted with only 0805 thick-film capacitors. Also component and PWB pad platings were included. The Ag-adhesive system needed 29% less material and 33% less energy for the interconnection process. However, the metal production energy was 42% more for Ag adhesive than solder. Leaching of metals was reported for TCLP at different pH values, and with a high likelihood that Pb is more prone to leach than Sn and Ag for pH 2 and 3.

Which main conclusions can be drawn from published work on environmental assessment of IM? The major conclusions regarding Pb-free solders compared to conventional 63Sn–37Pb are:

- The perception of hazards connected to the end-of-life phase is much higher for 63Sn–37Pb.
- As for overall environmental weighting methods such as EDIP, LIME, EPS, and Eco-Indicator '99, for which the resource consumption of Sn and Ag have

relatively high weighting indices, the Ag-containing Pb-free solders as a rule score higher than 63Sn–37Pb.

- For global warming the electricity usage in the reflow oven is the most important flow.

It has been awkward for the research community to establish whether conductive adhesives used in identical applications as solder pastes, are better or worse than the latter.

It has been argued that it is crucial by which routes the metals used in the materials are produced.

Another controversial issue is how much of the metals, especially Pb and Ag, will actually leach to nature from e-waste sent for long-term landfill and from the e-waste illegally dumped.

3.5.4 The Need for the Present Research

The following questions have been addressed to fill the gaps indicated in the literature.

Questions 2, 4–6, and 8 are related to the first overall hypothesis in Sect. 1.2, and questions 1–3 are related to the second.

Question number 7 is of complementary nature.

1. What is the consequential change in CO_2 emissions when a change between 95.5Sn–3.8Ag–0.7Cu and 62Sn–36Pb–2Ag solder paste is made?
2. What is the consequential change in LIME score when a change between 95.5Sn–3.8Ag–0.7Cu and 63Sn–37Pb solder paste is made?
3. What is the consequential change in Air and Water toxicity indicator scores when a change between 95.5S–3.8Ag–0.7Cu and 63Sn–37Pb solder paste is made?
4. What are the environmental implications, as estimated by CO_2 and LIME scores, of a shift from 95.5Sn–3.8Ag–0.7C to Ag-free solder pastes?
5. Using LIME, what are the life cycle differences between Sn and Au/Ni/Pd surface pad and terminal platings?
6. Would the availability of primary and more unit process data (Ag, electricity) change the LCA understanding of high-temperature adhesive/solder comparisons?
7. What are the differences in hidden material flows, so-called ore-TMR, for high-temperature adhesive/solder comparisons?
8. Using LIME, what is the difference between board-level LCA scores and scaled-up LCA scores reflecting total material usages?

References

1. Wackernagel M, Schulz NB, Deumling D et al (2002) Tracking the ecological overshoot of the human economy. Proc Int Acad Sci USA 99:9266–9271
2. Guinee JB (2002) Handbook on life cycle assessment: operational guide to the ISO Standards (eco-efficiency in industry and science), Kluwer, Dordrect
3. ISO (2006a) ISO 14040: environmental management–life cycle assessment-principles and framework. ISO 14040:2006(E)
4. Heijungs R (1994) A generic method for the identification of options for cleaner products. Ecol Econ 10:69–81
5. Sakai S, Yokoyama K (2002) Formulation of sensitivity analysis in life cycle assessment. Clean Technol Environ Policy 4:247–257
6. Takahashi K, Dodbiba G, Takeda T et al (2007) Assessing different recycling options for plastic wastes from discarded mobile phones in the context of LCA. Resour Process 54:29–34
7. Andrae ASG, Itsubo N, Yamaguchi H et al (2008) Life cycle assessment of high-temperature Japanese conductive adhesives. Environ Sci Technol 42:3084–3089
8. Heijungs R (2005) On the use of units in LCA. Int J LCA 10:173–176
9. Lin M, Zhang S, Chen Y (2005) Distance-to-target weighting in life cycle impact assessment based on Chinese environmental policy for the period 1995–2005. Int J LCA 10:393–398
10. Kim S, Hwang T, Lee KM (1997) Allocation for cascade recycling system. Int J LCA 2:217–222
11. Werner F, Richter K (2000) Economic allocation in LCA: a case study about aluminium window frames. Int J LCA 5:79–83
12. Reap J, Roman F, Duncan S et al (2008) A survey of unresolved problems in life cycle assessment–Part 1: goal and scope and inventory analysis. Int J LCA 13:290–300
13. Reap J, Roman F, Duncan S et al (2008) A survey of unresolved problems in life cycle assessment–Part 2: impact assessment and interpretation. Int J LCA 13:374–388
14. Yung WKC, Chan HK, Choi ACK, et al (2008) An environmental assessment framework with respect to the requirements of energy-using products directive. Proc Inst Mech Eng B-J Eng Manuf 222:643–651
15. Holland MR (1995) Assessment of the economic costs of damage caused by air-pollution. Water Air Soil Pollut 85:2583–2588
16. Rabl A, Spadaro JV (2000) Public health impact of air pollution and implications for the energy system. Ann Rev Energy Environ 25:601–627
17. Zhang M, Song Y, Cai X et al (2007) Economic assessment of the health effects related to particulate matter pollution in 111 Chinese cities by using economic burden of disease analysis. J Environ Manage 88:947–954
18. Greenpeace (2008) The true costs of coal
19. Jalal KF, Rogers PP (2002) Measuring environmental performance in Asia. Ecol Indic 2:39–59
20. Kitou E, Horvath A (2008) External air pollution costs of telework. Int J LCA 13:155–165
21. Japan Environmental Management Association for Industry (2005) LIME-LCA. ISBN 4-914953-96-X. In Japanese
22. Lee SY, Itsubo N, Motoshita M et al (2008) Development of LCCBA for eco-design in the electronics industry. Proc Electron Goes Green, Sept 7–10, Berlin, Germany, 479–482
23. Camargo J, Alonso A (2006) Ecological and toxicological effects of inorganic nitrogen pollution in aquatic ecosystems: a global assessment. Environ Int 32:831–849
24. Bengtsson M, Steen B (2000) Weighting in LCA-approaches and applications. Environmental Progress 19:101–109
25. Goedkoop M, Hofstetter P, Müller-Wenk R et al (1998) The eco-indicator 1998 explained. Int J LCA 6:352–360

26. Low MK (1997) Design and financial aspects of the end-of-life management of telecommunications products. PhD thesis, Loughborough University, UK, pp. 58–66
27. Vignes RP (2001) Limited life-cycle analysis for environmental decision-making. Chem Eng Prog 97:40–54
28. Verhoef EV, Dijkema GPJ and Reuter MA (2004) Process knowledge, system dynamics and metal ecology. J Ind Ecol 8:23–43
29. Reuter MA, Verhoef EV (2004) A dynamic model for the assessment of the replacement of lead in solders. J Electron Mater 33:1567–1580
30. Itsubo N, Noh J, Inaba A (2004) Life cycle impact assessment of lead-free solder considering the trade-off relationship between toxic impact and other impact categories. Mater Trans 45:3194–3200
31. US Environmental Protection Agency (2005) Solders in electronics: a life cycle assessment. EPA, Cincinnati, OH
32. PRe' Consultants B.V. (2005) LCA comparison of alternative soldering techniques
33. Ekvall T, Andrae ASG (2006) Attributional and consequential environmental assessment of the shift to lead-free solders. Int J LCA 11:344–353
34. http://www.ecoinvent.org/
35. Segerberg T, Hedemalm P (1996) Life cycle assessment of tin-lead solder and silver-epoxy conductive adhesive. Proc IEEE Int Symp Electron Environ, 6–8 May, Dallas, TX, pp 64–65
36. Liu J, Tolvgard A, Malmodin J et al (1999) A reliable and environmentally friendly packaging technology-flip-chip joining using anisotropically conductive adhesive. IEEE Trans Electron Packag Manuf 22:186–190
37. Dansk Toksikologi Center (2000) Life cycle assessment of electrically conductive adhesive vs. traditional tin/lead solder. Working report for the Kamille II project, Horsholm, Denmark
38. Andrae ASG, Itsubo N, Yamaguchi H et al (2007) Screening life cycle assessment of silver-based conductive adhesive vs. lead-based solder and plating materials. Mater Trans 48:2212–2218
39. Andrae ASG, Itsubo N, Yamaguchi H et al (2008) Life cycle assessment of high-temperature Japanese conductive adhesives. Environ Sci Technol 42:3084–3089
40. Nakamura S, Murakami S, Nakajima K et al (2008) Hybrid input–output approach to metal production and its application to the introduction of lead-free solders. Environ Sci Technol 42:3843–3848
41. Fuse M, Kashima S (2008) Evaluation method of automobile recycling systems for Asia considering international material cycles: application to Japan and Thailand. J Mater Cycles Waste Manag 10:153–164
42. Fukuda Y, Pecht MG, Fukuda K (2003) Lead-free soldering in the Japanese electronics industry. IEEE Trans Comp, Pack Manuf Technol 26:616–624
43. Schischke K, Griese H, Muller J et al (2003) Cost-effectiveness and environmental aspects of flip chip bumping for system integration. Microsys Technol 9:324–330
44. Wu BY, Chan YC, Middendorf A et al (2008) Assessment of toxicity potential of metallic elements in discarded electronics: a case study of mobile phones in China. J Environ Sci 20:1403–1408
45. Andrae ASG, Andersson DR, Ramberg C et al (2005) Application of simplified methods for partial environmental assessment of microelectronics soldering materials and processes in small and medium sized enterprises. Gate to EHS.
DOI: http://dx.doi.org/10.1065/ehs2005.05.XXX
46. Bergendahl CG, Lichtenvort K, Johansson G et al (2005) Environmental and economic implications of a shift to halogen-free printed wiring boards. Circuit World 31: 26–31
47. Park PJ, Lee KM, Wimmer W (2006) Development of an environmental assessment method for consumer electronics by combining top-down and bottom-up approaches. Int J LCA 11:254–264

48. Zhou X, Schoenung JM (2007) An integrated impact assessment and weighting methodology: evaluation of the environmental consequences of computer display technology substitution. J Environ Manag 83:1–24
49. Zhou X, Schoenung JM (2008) An integrated impact assessment and weighting methodology: evaluation of the environmental consequences of lead-free solder alternatives. Proc IEEE Electron Environ, San Fransisco, CA, May 19–22
50. Griese H, Muller J, Reichl H et al (2000) Environmental assessment of lead free interconnection systems. Proc Symp Lead Free Intercon Technol, SMTA, Boston, MA, June 13–14.
51. Hamano H, Suga T, Okamoto M et al (2001) Environmental impact evaluation for the full life cycle of products using Pb-free solders. Proc Second Int Symp Environ Conscious Design Inverse Manuf, 11–15 Dec., Tokyo, Japan, 1079–1083
52. Turbini LJ, Munie GC, Bernier D et al (2001) Examining the environmental impact of lead-free soldering alternatives. IEEE Trans Electron Packag Manuf 24:4–9
53. Graedel TE (2002) Material substitution: a resource supply perspective. Resour Conserv Recycl 34:107–115
54. Ku A, Ogunseitan O, Saphores J et al (2003) Lead-free solders: issues of toxicity, availability and impacts of extraction. Proc 53rd Electron Comp Technol Conf, New Orleans, LA, 27–30 May, 47–53
55. Mihaescu R, Carstea H (2003) Ecological technologies for soldering without lead in electronic and telecommunications – today and tomorrow. Proc 26th Int Spring Seminar on Electron Technol: Integrated Managem of Electron Mater Prod, 8–11 May, High Tatras, Slovak Republic, 232–234
56. Deubzer O (2007) Explorative study into the sustainable use and substitution of soldering metals in electronics. PhD thesis, Delft University, Delft, the Netherlands
57. Conway PP (1993) An environmental comparison of solder and conductive adhesives for SMT interconnect. Proc 15th IEEE/CHMT Intl Electron Man Technol Symp, 4–6 Oct, pp. 171–176.
58. Muller J, Griese H, Reichl H (1999) Reduced environmental impacts by lead free electronic assemblies? Int Summit Lead-Free Electron Assembl. Anhang A IPCWorks'99 in Minneapolis, U.S.A., 26–2 Oct.

Chapter 4
Methodology

Abstract This chapter describes in detail how LCIA calculations can be performed in practice to compare interconnection materials. Several software programs are discussed with special emphasis on the Excel-based software EMLCA. One possible methodology for performing an LCIA of interconnection materials is presented. Concerning consequential LCA, a procedure for identifying marginal consumers is proposed. This text makes a summary of the important issues regarding the remaining needs for carrying out a successful consequential LCA in practice. The effect on the markets of Pb, Ag, Sn, and Bi as a result of the Pb-free shift is explored.

4.1 Software

For life cycle assessments several well-known software programs exist [1]. Software is necessary to carry out LCA in order to reduce time for gathering information [1].

Recent LCA free-ware programs include *openLCA* [2]. *openLCA* includes a converter which can translate different LCI database formats into one another.

4.1.1 SimaPro

SimaPro is a Dutch LCA program developed by PRe' Consultants. Its strengths lie within the transparency by which data are viewed and results presented.

Moreover, in opposite to other LCA programs, it can correctly calculate the probability (see Sect. 4.7.1) of one alternative being better than another.

SimaPro is the one of the most widely used LCA software worldwide with users in 60 countries.

Truth be told, SimaPro is considered to be a very user-friendly and reliable tool with constant updates and a professional organization behind it, PRe' Consultants. Anything regarding LCA calculations can be carried out.

An advantage of the tool is the interactive results analysis, which gives a very high transparency, as well as the ability to calculate "looped" systems.

An additional benefit for multinational companies is that SimaPro can be used by several users at the same time irrespective of geographical location.
Annually, results from numerous LCA case studies are published where SimaPro has been used.

One of SimaPro's disadvantages is that so far the life cycle impact assessment uncertainty cannot be included in the total uncertainty analysis (only inventory uncertainty can be modeled using Monte Carlo simulation).

PRe' Consultants are intending to include the LCIA uncertainty in future updates of their LCA software. In Sect. 4.7 software capabilities to calculate uncertainty are further elaborated.

Another advantage is that PRe' collaborates with the Swiss *ecoinvent* database famous for its vast contents of quality life cycle inventory (LCI) data of background processes.

The recently released version 2 of the database has life cycle profiles of nearly 4,000 processes and materials, facilitating, e.g., screening LCAs.

Except for LCI data the database also comes with many impact assessment methods (such as Eco-Indicator '99 and LIME).

4.1.2 EcoLab

EcoLab is a Swedish LCA program which has been used mainly within the Swedish industry. Andrae used EcoLab for exploring the applicability of LCA in microelectronics packaging [3].

EcoLab's main advantage is its relational database making reuse of components/modules very easy.

The single Oracle® database contains all activities, flowchart scenarios, processes, life cycle scenarios, and substance nomenclature and there are no "library projects" or "projects" as in SimaPro. Moreover, for each analysis the results of all impact and weighting methods can be visualized.

Below in Fig. 4.1 an example is given of an EcoLab process tree for a solder paste. Figure 4.2 shows a weighting result of three interconnection materials within CMLCA. Neither closed loops nor uncertainty analysis can be performed in EcoLab which is a huge disadvantage compared to, e.g., EMLCA and SimaPro.

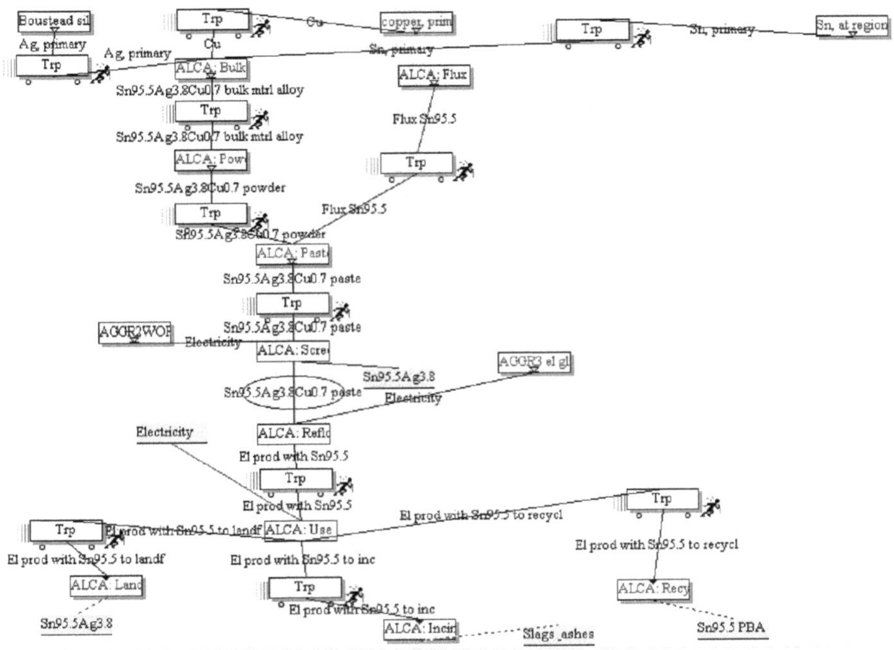

Fig. 4.1 A process tree for the life cycle of a solder paste as shown in EcoLab LCA software

4.1.3 Chain Management by Life Cycle Assessment

Chain Management by Life Cycle Assessment (CMLCA) is a Dutch LCA freeware program developed by the distinguished environmental systems analysis group at Leiden University in the Netherlands.

Its main strength compared to other LCA software is that the uncertainty of the LCIA part can be included. Below in Tables 4.1 and 4.2 the unique CMLCA feature of including the uncertainty for characterization and weighting is shown.

L(1.74) for nitric oxide means in CMLCA a lognormal distribution corresponding to a coefficient of variation of 443% as shown in Table 3.2.

Table 4.1 The possibility of including uncertainty of characterization or damage factors within CMLCA

	Environmental emissions			
Label	Name	Value	Unit	Uncertainty
[E50]	Nitrogen oxides[air]	1.46×10^{-8}	DALY/g	L(1.74)
[E53]	Part. Be. 2.5 and 10[air]	2.53×10^{-7}	DALY/g	L(1.45)
[E54]	Part. Bi. 10.[air]	2.53×10^{-7}	DALY/g	L(1.57)
[E55]	Part. L. 2.5[air]	4.16×10^{-7}	DALY/g	L(1.37)
[E67]	Sulf. Oxi.[air]	1.05×10^{-7}	DALY/g	L(1.06)
[E68]	Sulfur dioxide[air]	1.05×10^{-7}	DALY/g	L(1.06)
[E96]	Nitric oxide[air]	1.46×10^{-8}	DALY/g	L(1.74)
[E98]	Particulates[air]	2.53×10^{-7}	DALY/g	L(1.4)
[E99]	Particulates, < 10 um[air]	2.53×10^{-7}	DALY/g	L(1.4)
[E100]	Particulates, < 10 um (mobile)[air]	1.14×10^{-6}	DALY/g	L(1.4)
[E101]	Particulates, < 10 um (stationary)[air]	2.53×10^{-7}	DALY/g	L(1.4)
[E102]	Particulates, > 10 um (process)[air]	2.53×10^{-7}	DALY/g	L(1.4)
[E103]	Particulates, SPM[air]	2.53×10^{-7}	DALY/g	L(1.4)

Table 4.2 The possibility of including uncertainty of weighting factors within CMLCA

		Category weights		
Label	Name	Value	Unit	Uncertainty
[C1]	Global warming (HH)	9.70×10^{6}	–	L(0.452)
[C2]	Global warming (SA)	1	–	L(0.287)
[C3]	Ozone layer depletion (HH)	9.70×10^{6}	–	L(0.452)
[C4]	Ozone layer depletion (SA)	1	–	L(0.287)
[C5]	Ozone layer depletion (PP)	20.2	–	L(0.416)
[C6]	Photochemical oxidants (HH)	9.70×10^{6}	–	L(0.452)
[C7]	Photochemical oxidants (SA)	1	–	L(0.287)
[C8]	Photochemical oxidants (PP)	20.2	–	L(0.416)
[C9]	Acidification (SA)	1	–	L(0.287)
[C10]	Acidification (PP)	20.2	–	L(0.416)
[C11]	Ecotoxicity (BD)	4.80×10^{12}	–	L(0.327)
[C12]	Air pollution (HH)	9.70×10^{6}	–	L(0.452)
[C13]	Human toxicity (HH)	9.70×10^{6}	–	L(0.452)
[C14]	Eutrophication (SA)	1	–	L(0.287)
[C15]	Resource consumption (SA)	1	–	L(0.287)
[C16]	Resource consumption (PP)	20.2	–	L(0.416)
[C17]	Resource consumption (BD)	4.80×10^{12}	–	L(0.327)
[C18]	Waste (BD)	4.80×10^{12}	–	L(0.327)
[C19]	Waste (PP)	20.2	–	L(0.416)

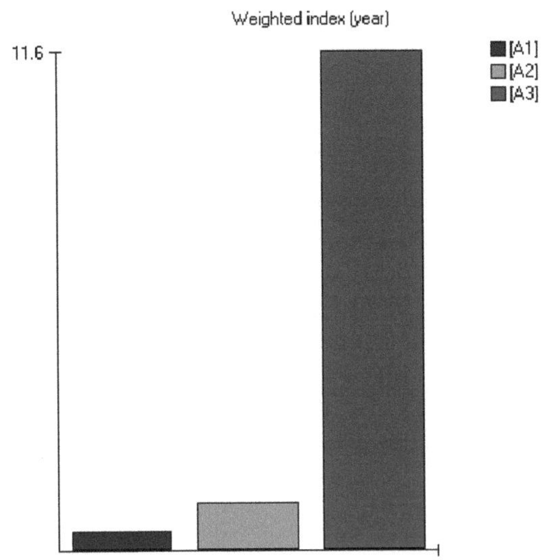

Fig. 4.2 Comparison from CMLCA of three interconnection materials using weighting

4.1.4 Excel Management by LCA

In 2006 Excel Management by LCA (EMLCA) was developed by the Chinese Ph.D. student Weizhe Lu at the University of Tokyo [4]. EMLCA is a very pedagogic (especially for explaining the effects of recycling) LCA free-ware program based on Microsoft Excel®.

Below in Table 4.3 is shown the ease by which EMLCA displays a life cycle of an alarm clock described by Sullivan [5].

In Tables 4.3–4.5 the number to the right of PROCESS indicates the number of processes, and the number below each process name indicates the number of flows in each process.

In order to obtain the LCI results, the user can copy Tables 4.3–4.5 into EMLCA and its "Sheet" for "Original data," then "Initialize data", then "Calculate Environmental Loads."

Table 4.3 EMLCA calculation scheme for an alarm clock using DIP technology and 63Sn–37Pb alloy

PROCESS	10		Mean	Standard deviation
electricity, hard coal, at power plant operation				
5				
F	electricity, hard coal, at power plant	kWh	1	
I	hard coal, burned in power plant	MJ	−10.1	
L	Pb, air	kg	6.1E–08	
L	Pb, soil	kg	7.44E–11	
L	Pb, water	kg	4.59E–08	
hard coal burning				
5				
F	hard coal, burned in power plant	MJ	1	
I	hard coal supply mix	kg	–0.0453	
L	Pb, air	kg	6.32E–08	
L	Coal required	kg	0.0453	
O	Coal ashes for landfill	kg	0.00715	
hard coal supply mix				
2				
F	hard coal supply mix	kg	1	
I	hard coal, at mine	kg	−1.03	
hard coal mining				
2				
F	hard coal, at mine	kg	1	
L	Coal, hard, unspecified, in ground	kg	1.21	
Pb production				
5				
F	Pb	kg	1	
L	Coal, hard, unspecified, in ground	kg	0.22624	
L	Pb, air	kg	8.51E–05	
L	Pb, soil	kg	1.35E–08	
L	Pb, water	kg	4.88E–06	
Alarm clock production				
2				
F	Alarm clock	p	1	
I	Pb	kg	–0.00083	

Table 4.3 continued

PROCESS	10		Mean	Standard deviation
Alarm clock usage 10 years				
5				
F	Alarm clock use	p	1	
I	Alarm clock	p	–1	
I	electricity, hard coal, at power plant	kWh	–131.4	
O	Incineration of Alarm clock	p	0.1	
O	Landfill of Alarm clock	p	0.9	
Alarm clock incineration process				
4				
F	Incineration of Alarm clock	p	–1	
L	Pb, air	kg	4.15E–05	
L	Pb, soil	kg	0.000394	
L	Pb, water	kg	0.000394	3.94E–5
Alarm clock landfill process				
5				
F	Landfill of Alarm clock	p	–1	
I	electricity, hard coal, at power plant	kWh	–0.00144	
I	Alarm clock to landfill	p	–1	
L	Pb, soil	kg	0.000415	
L	Pb, water	kg	0.000415	4.15E–4
Coal ash landfill process				
2				
F	Coal ashes for landfill	kg	–1	
L	Pb, water	kg	0.00004	
BOUNDARY	1			
B	Alarm clock use	p	1000000	

Table 4.4 EMLCA calculation scheme for an alarm clock using current packaging technology and 63Sn–37Pb alloy

PROCESS	10		Mean	Standard deviation
electricity, hard coal, at power plant operation				
5				
F	electricity, hard coal, at power plant	kWh	1	
I	hard coal, burned in power plant	MJ	–10.1	
L	Pb, air	kg	6.1E–08	
L	Pb, soil	kg	7.44E–11	
L	Pb, water	kg	4.59E–08	4.59e–9
hard coal burning				
5				
F	hard coal, burned in power plant	MJ	1	
I	hard coal supply mix	kg	–0.0453	
L	Pb, air	kg	6.32E–08	
L	Coal required	kg	0.0453	
O	Coal ashes for landfill	kg	0.00715	
hard coal supply mix				
2				
F	hard coal supply mix	kg	1	
I	hard coal, at mine	kg	–1.03	
hard coal mining				
2				
F	hard coal, at mine	kg	1	
L	Coal, hard, unspecified, in ground	kg	1.21	
Pb production				
5				
F	Pb	kg	1	
L	Coal, hard, unspecified, in ground	kg	0.22624	
L	Pb, air	kg	8.51E–05	
L	Pb, soil	kg	1.35E–08	
L	Pb, water	kg	4.88E–06	4.88E–7
Alarm clock production				
2				
F	Alarm clock	p	1	
I	Pb	kg	–3E–05	

Table 4.4 continued

PROCESS	10		Mean	Standard deviation
Alarm clock usage 10 years				
5				
F	Alarm clock use	p	1	
I	Alarm clock	p	–1	
I	electricity, hard coal, at power plant	kWh	–78.84	
O	Incineration of Alarm clock	p	0.1	
O	Landfill of Alarm clock	p	0.9	
Alarm clock incineration process				
4				
F	Incineration of Alarm clock	p	–1	
L	Pb, air	kg	1.48E–06	
L	Pb, soil	kg	1.41E–05	
L	Pb, water	kg	1.41E–05	1.41E–6
Alarm clock landfill process				
5				
F	Landfill of Alarm clock	p	–1	
I	electricity, hard coal, at power plant	kWh	–0.00144	
I	Alarm clock to landfill	p	–1	
L	Pb, soil	kg	1.48E–05	
L	Pb, water	kg	1.48E–05	1.48E–6
Coal ash landfill process				
2				
F	Coal ashes for landfill	kg	–1	
L	Pb, water	kg	0.00004	4E–6
BOUNDARY	1			
B	Alarm clock use	p	1000000	

Table 4.5 EMLCA calculation scheme for an alarm clock using current packaging technology and Pb-free alloy

PROCESS	10		Mean	Standard deviation
electricity, hard coal, at power plant operation				
5				
F	electricity, hard coal, at power plant	kWh	1	
I	hard coal, burned in power plant	MJ	–10.1	
L	Pb, air	kg	6.1E–08	
L	Pb, soil	kg	7.44E–11	
L	Pb, water	kg	4.59E–08	
hard coal burning				
5				
F	hard coal, burned in power plant	MJ	1	
I	hard coal supply mix	kg	–0.0453	
L	Pb, air	kg	6.32E–08	
L	Coal required	kg	0.0453	
O	Coal ashes for landfill	kg	0.00715	
hard coal supply mix				
2				
F	hard coal supply mix	kg	1	
I	hard coal, at mine	kg	–1.03	
hard coal mining				
2				
F	hard coal, at mine	kg	1	
L	Coal, hard, unspecified, in ground	kg	1.21	
Pb free alloy production				
5				
F	Pb free alloy	kg	1	
L	Coal, hard, unspecified, in ground	kg	2.9436	
L	Pb, air	kg	0.000257	
L	Pb, soil	kg	6.25E–08	
L	Pb, water	kg	4.22E–05	
Alarm clock production				
2				
F	Alarm clock	p	1	
I	Pb free alloy	kg	–7E–05	

Table 4.5 continued

PROCESS	10		Mean	Standard deviation
Alarm clock usage 10 years				
5				
F	Alarm clock use	p	1	
I	Alarm clock	p	–1	
I	electricity, hard coal, at power plant	kWh	–78.84	
O	Incineration of Alarm clock	p	0.1	
O	Landfill of Alarm clock	p	0.9	
Alarm clock incineration process				
4				
F	Incineration of Alarm clock	p	–1	
L	Pb, air	kg		
L	Pb, soil	kg		
L	Pb, water	kg		
Alarm clock landfill process				
5				
F	Landfill of Alarm clock	p	–1	
I	electricity, hard coal, at power plant	kWh	–0.00144	
I	Alarm clock to landfill	p	–1	
L	Pb, soil	kg		
L	Pb, water	kg		
Coal ash landfill process				
2				
F	Coal ashes for landfill	kg	–1	
L	Pb, water	kg	0.00004	
BOUNDARY	1			
B	Alarm clock use	p	1000000	

In Fig. 4.3 is displayed a graph of the connections in Table 4.5.

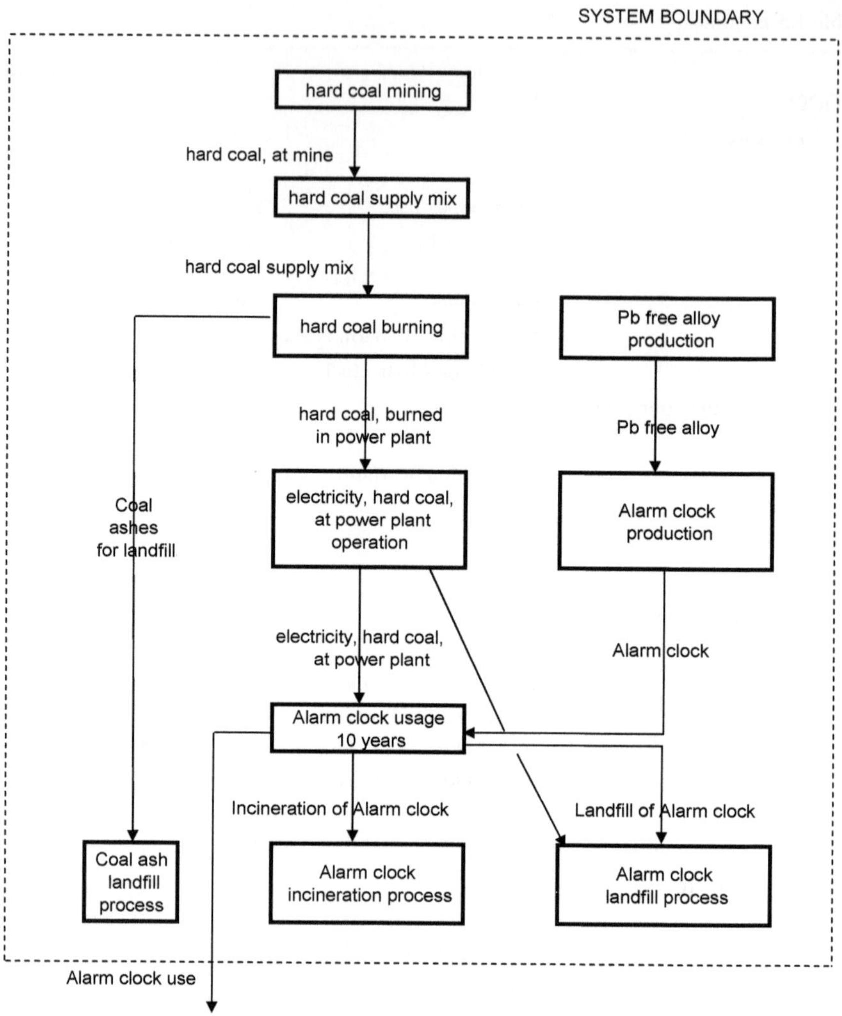

Fig. 4.3 Graphical illustration of Table 4.5

Using Eq. 3.1, "B" in Tables 4.3–4.5 is the f_u, "F" is f, "L" is B, and "I" and "O" are A.

As mentioned in Chap. 1, Sullivan [5] argued that the Pb content of the coal used in the life cycle of electronics is higher than the Pb content of the device. EMLCA and other LCA tools can help keeping track of such calculations.

It seems like the old-fashioned dual inline package (DIP) technology from 1994 would mean a larger share of the Pb emissions from the landfill of the alarm clock than the coal ash Pb emissions.

Present "re-eco-designed" alarm clocks have a standby power usage of less than 1W, less Pb per joint and/or Pb-free design. The differences are shown in Fig. 4.4.

A hypothesis is that coal ash landfilling has gradually become the major source for potential Pb emissions in the life cycle of electronic devices.

Naturally the user can change to any modern electronic device and put in measured data for all fields in Tables 4.3–4.5.

The uncertainty is included in EMLCA by adding the value corresponding to two standard deviations (2σ) in a 95% confidence interval. For example for "Pb, water" in Table 4.3 the total uncertainty, Sd, is 37.55 kg, i.e., the CoV is 2.3%.

Table 4.6 Summary of uncertainties for one million alarm clocks Design year 1994

Load	Pb, air	Pb, soil	Pb, water	Coal required	Coal, hard, unspecified, in ground
Mean	96.10708	412.4373	798.0286	60120033	74927785
Sd(A)	0	0	0	0	0
Sd(B)	0	0	37.55724	0	0
Sd	0	0	37.55724	0	0

For the new alarm clock shown in Table 4.4 the uncertainty is 22.82 kg and CoV 4.6% (Table 4.6).

Table 4.7 Summary of uncertainties for one millions alarm clocks design year 2008

Load	Pb, air	Pb, soil	Pb, water	Coal required	Coal, hard, unspecified, in ground
Mean	55.28581	14.73187	246.0859	36072256	44956859
Sd(A)	0	0	0	0	0
Sd(B)	0	0	22.81632	0	0
Sd	0	0	22.81632	0	0

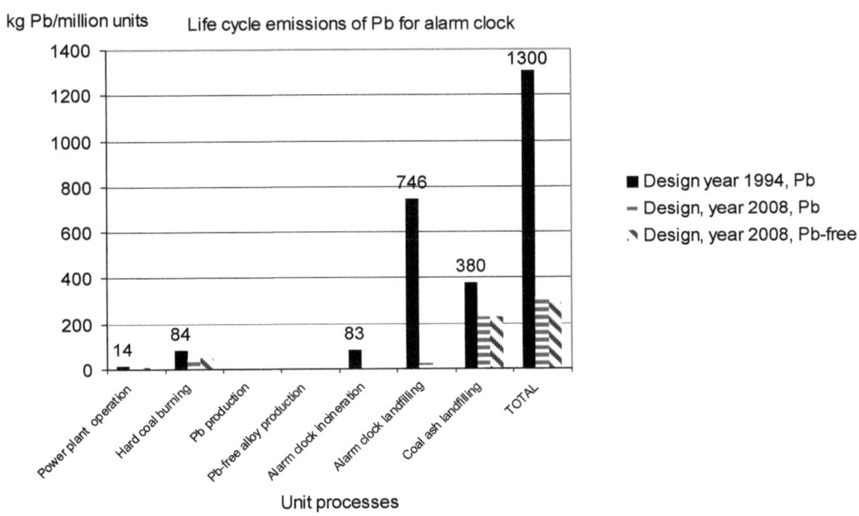

Fig. 4.4 Life cycle emissions of Pb for an alarm clock

This information can be used to determine the probability of the Pb emissions to water being higher for either alarm clock.

4.2 Databases

The backbone of all calculations is the background data by which the current environmental loadings are estimated.
If inventory data for electricity and basic materials, both of which are used in many products, is already there, it is possible for, e.g., a company to reduce the analysis time.

As *imagined* by most practitioners, the lack of reliable databases is the main problem of the practical application of LCA. Another problem is the difficult choice of the most reliable impact assessment method. It is expected that extensive reporting legislation in combination with standardization will be the most fruitful ways for creation of robust LCI databases.

The most comprehensive and transparent LCI database for Europe is *ecoinvent* [6]. Several national LCI databases are updated and created every year, e.g., for Japan and Australia.

The need encouraged the European Commission to start a project aiming at "globally consistent, quality assured data and method recommendations for LCA", an international reference life cycle data platform based on quality-assured LCA studies [7].

Occasionally, the existing databases are specific enough to give indications of the magnitude of the *background system* (e.g., basic raw material production and energy conversions).

For the *foreground system* the practitioner should collect primary data such as material and energy usages.

4.3 Methodology for Consequential Process-sum LCIA Using LIME and Uncertainty Analysis

One possible methodology for performing an LCIA of interconnection materials (IM), e.g., wave solder, solder paste, etc., according to the 31 following steps, was developed based on the practical experience from previous research [3]:

1. Define the goal and scope of the study, including defining the functional unit and system boundaries.
2. Break down the IM into a structure which describes how the different parts fit together, and identify the materials list.

3. Design a detailed flow diagram of the product system where the manufacturing, use and end-of-life stages shall be modeled. A flow diagram tree is displayed in Fig. 4.5 for Sn–Pb solder paste.
4. Define the material content declarations for the IMs in the LCA software (e.g., SimaPro).

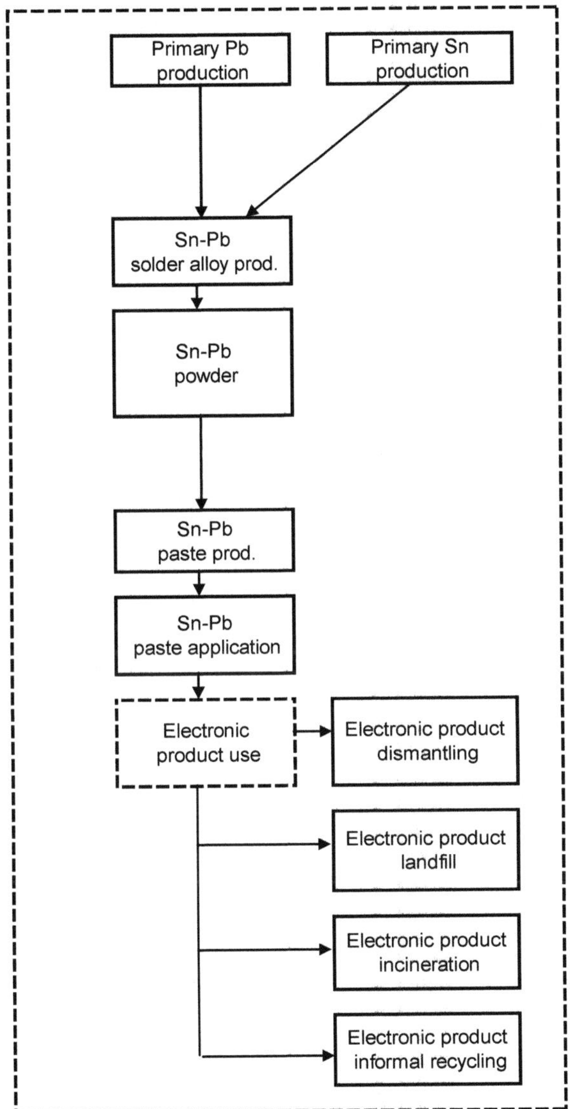

Fig. 4.5 Flow diagram in ALCA for Sn–Pb solder paste

5. Determine the transportation distances and modes for the main transportations: i.e., from the alloy manufacturers to the manufacturing of the IM (PM), and from the PM to the solder paste application.
6. Find out which is the most probable end-of-life treatment method for the IM.
7. Find out the total IM amount produced during, e.g., one year and what proportion of the IM material content is recycled.
8. Find out the electricity usage during the use phase (stencil printing, component mounting, and reflow process) of the IM product life. Do this also for maintenance or installation if these activities are relevant or suspected to have a significant load.
9. When collecting LCI data (material and energy flows, resource consumptions and emissions), choose from the sub-material manufacturers from which larger quantities are bought.
10. Outline a questionnaire in which sub-material manufacturers are asked for total consumption figures and total production figures for the specific factory, for transport to the factory and transport to waste treatment. Require the mass share of the sub-material in comparison to other outputs.
11. For those processes and activities which are judged as having inappropriate matching with the databases in SimaPro, other data should be used. Appropriate modules for LCI data for chemicals and materials are preferably obtained from literature and imported to SimaPro.
12. Based on Fig. 4.5, make processes which are used as inputs and outputs of each other. In order to keep track of the "material content" of the IM, add flows to the Resources input field (of SimaPro) such as "MD, Sn" and "MD, Ag." The purpose of adding these "dummy" flows are to keep track of the magnitude of materials which will enter the end-of-life phase.
13. Find out the appropriate electricity mix for the LCI models, raw materials, and everywhere else where electricity is used. The electricity mix used, e.g., in the ecoinvent database might not reflect the conditions under study whereby adaption will be necessary.
14. If feasible, find out how much electricity and other energies are used in the offices for developing the studied IM.
15. Find out how much energy and materials are totally consumed in, e.g., the IM factory per year.
16. Use the total material declaration, defined in step 4, as the input for end-of-life.
17. Create different recycling scenarios for the electronic device, e.g., landfill, incineration, etc. Possibly, the method presented by Huisman can be used, to choose the most eco-efficient recycling scenario for the IM [8].
18. Assume recycling percentages for different metals.
19. Set the base flow for the functional unit, f_u, e.g., amount IM needed to cover a specific pad area.
20. Let the LCA software calculate the LCI result.
21. Check that the nomenclature of characterization and weighting indices for different materials within the LCA software match the nomenclature of LCI re-

sult. The material content of "MD, X" is compared to the material content defined in step 4.
22. Analyze the characterization results.
23. Analyze the weighting results.
24. Normalize the results, i.e., compare the characterization/weighting results with the total characterization/weighting results for the region, which is within the system boundary, e.g., Europe or UK. By this, one can understand the relative importance of the different environmental effects caused by the IM.
25. Perform contribution analysis.
26. Perform gravity analysis.
27. Perform uncertainty analysis.
28. Perform a sensitivity analysis, using *CMLCA*[3], on the results to see if the results change significantly due to changes in input variables. E.g., put in higher electricity usage in the use phase and see how its relationship to the other life cycle phases changes.
29. Identify the significant environmental aspects of the IM. Aspects could be electricity usage in the reflow process, toxic substance emissions in the unregulated end-of-life treatment, etc.
30. Evaluate the completeness, sensitivity and consistency of the LCA study.
31. Start the CLCA process described in Sect. 4.4.

4.4 Consequential LCI

Consequential LCA (CLCA), to a higher extent reflecting the real world than the conventional attributional LCA (ALCA), is proposed to give more realistic information about the consequences of buying a product.

CLCA can be applied to identical systems as ALCA. In the literature ALCA is the most published approach by far, but some authors have tried (with more or less limited success) to apply the CLCA approach [9–13]. Table 4.8 shows some of the differences between ALCA and CLCA.

This text makes a summary of the important issues regarding the remaining needs for practically carrying out a successful consequential LCA.

One of the most urgent needs for a successful development of CLCA is the involvement of econometricians.

The most commonly used variant of LCA is the ALCA describing the environmental impact of the life cycle of a product using e.g., nanoparticles.

[3] CMLCA is likely the only LCA software which facilitates uncertainty and sensitivity analysis based on "all" elements of LCA.

Table 4.8 Differences between attributional and consequential LCA

	Attributional	Consequential
Advantages	Understanding of processes	Address the effect in relevant markets
	Knowledge of life cycle cost and eco-efficiency	Founded on national economic theories
	Based on average and historical data	
Drawbacks	Descriptive.	Physical reality is different from socio-economic thinking → failure to identify marginal technologies
	A limited system is accounted for → not able to assess environmental effects outside the chosen system	Rapid changes in technological level make assessment of consequences and monumental task
	High degree of uncertainty	High degree of uncertainty
		Lack of data

CLCA is another variant of LCA which aims at describing how the environmentally relevant physical flows to and from the technosphere will change in response to possible changes made within the life cycle of a product, e.g., increase the Sn and Ag content of a solder at the expense of Pb.

Earlier research has shown that CLCA is more suitable for assessments of technological shifts having a larger share of the world market than applications of limited importance.

ALCA typically advocates allocation and average data, whereas CLCA involves system expansion and marginal data.

Separating the LCA approaches into ALCA and CLCA moves the discussion further on what knowledge an LCA actually generates.

CLCA is not entirely new but is somewhat immature and unexplored, however, promising research is starting to appear. For instance, related to CLCI, researchers from United States explored the changes of demand of electricity caused by price changes of CO_2. The econometrically inspired paper illustrates the price elasticity of consumer demand for various CO_2 prices. The price elasticities of carbon dioxide were modeled using econometric techniques [14].

The possibilities of CLCA, for capturing consequential side-effects as a result of market shifts caused by changing from one technology to another, are hypothetically many.

In Fig. 4.6 the rough difference between the ALCA and the CLCA is shown.

By ALCA the potential impact of life cycle A (old technology) and life cycle B (new technology) are compared, and the practitioner decides which one is best based on the magnitude of the scores.

CLCA on the other hand attempts to quantify the *resulting total consequences* of a shift as the "consequences of changes within and outside life cycle A" subtracted from the "consequences of changes within and outside of life cycle B." The circles below in Fig. 4.6 represent the global environmental load. CLCA involves calculating own-price elasticities of supply and demand, and identifying marginal consumers and producers.

CLCA could be imagined as more difficult than ALCA as the systems investigated become more complex and because the marginal technology is often difficult to identify.

On the other hand, CLCA only requires data for the marginal technology and ALCA for larger product systems. There is no consensus on how to identify the marginal technology.

As an example of what type of questions CLCA helps to answer, an example of the global replacement of Pb in Pb-free solders is provided. For ALCA two or more LCAs were performed, one for the Pb solder and one or more for the Pb-free replacement(s). That is how the solder LCAs so far have been carried out and the unit operations commonly included are shown in Fig. 4.5. The results (e.g., CO_2 emissions) were compared by several authors but not much about the actual global (or national) consequences can theoretically be estimated.

From Fig. 4.5, representing a conventional ALCA, it is difficult to generate knowledge about *processes affected outside the life cycle* of the solder. The conventional ALCA compares, e.g., total CO_2 and Pb emissions (or estimation of the "total" environmental impact) of two or more life cycles to each other and also finds hot-spots within the life cycles.

However, the "real" consequence is that Pb will not be used in low-melting-point Sn–Pb solders but used in other products (by other consumers) instead. The CLCA, in opposition to the ALCA, includes as far as possible these other products' life cycles.

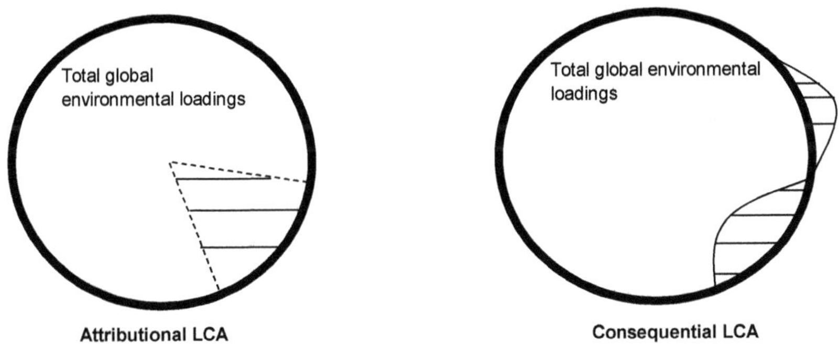

Fig. 4.6 The conceptual difference between ALCA and CLCA

An important tool used in CLCA is partial equilibrium modeling describing supply and demand of products. The tool is used to investigate how supply/demand changes by a change in price of the products, or by alterations on the supply or demand side. Partly the Pb price will change as a result of the ban of Pb in solders. (However, the Pb price change will likely be small as only 0.8wt% of the global Pb is used for solders.) Nevertheless, the *marginal* Pb consumers are the ones *most sensitive* to a Pb price change. Further in this chapter a six-step procedure is suggested to identify these consumers.

The consumption of Pb will obviously decrease in the solder life cycle and increase elsewhere as *the marginal Pb consumers* will act to consume more Pb, as they can afford that and will find it profitable to substitute what they currently use. The total Pb consumption might not increase, though.

The preliminary structure of the CLCI system of solder paste where the Pb market and Pb scrap market is included, is presented in Figs. 4.7 and 4.8.

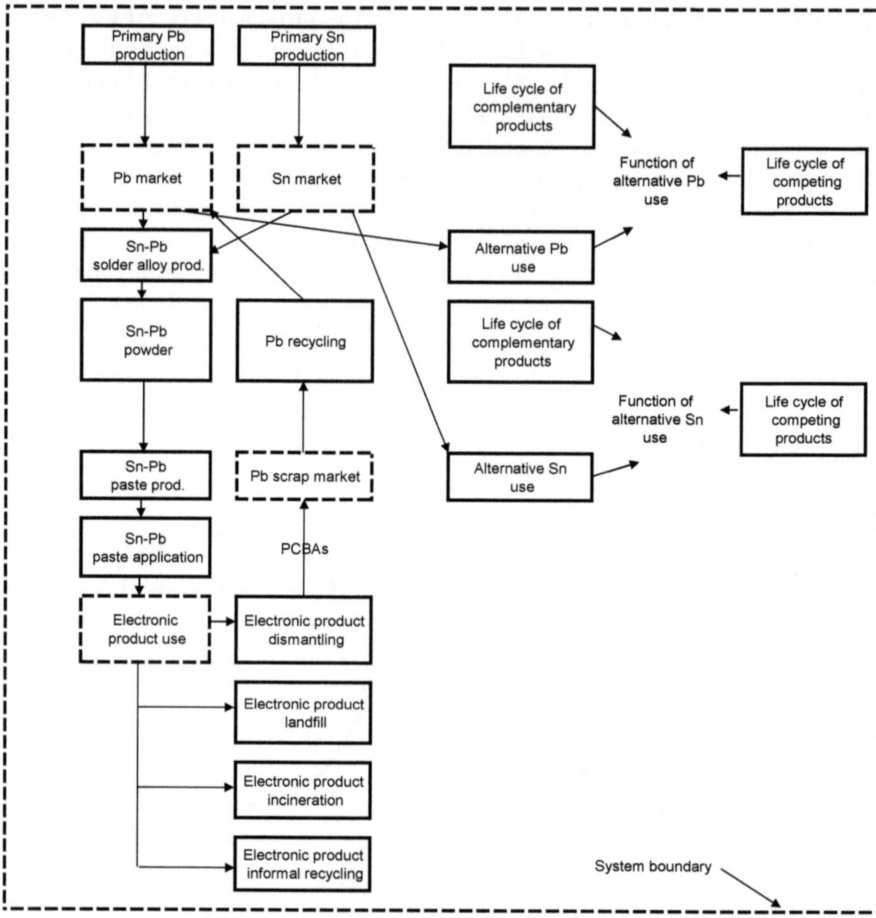

Fig. 4.7 The schematic CLCI model for a Sn–Pb solder including the Pb and Pb scrap markets

Global or local own-price elasticities (long-term or short-term) describe how the quantity supplied, x_s or demanded, x_d, will change when the price, y, changes. Below in Eqs. 4.1–4.3 are shown the mathematical definitions of k_s and k_d.

$$k_s = \frac{y}{x_s}\left(\frac{\partial x_s}{\partial y}\right) \tag{4.1}$$

$$k_d = \frac{y}{x_d}\left(\frac{\partial x_d}{\partial y}\right) \tag{4.2}$$

$$\frac{k_s}{k_\mathrm{d}} = \frac{\partial x_s}{\partial x_d} \tag{4.3}$$

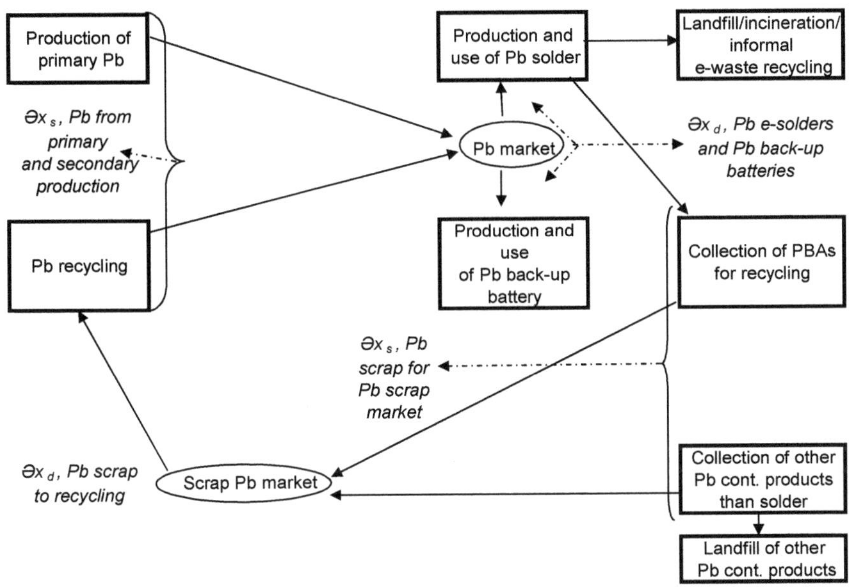

Fig. 4.8 A CLCA model for the Pb and scrap Pb markets after the shift to Pb-free solder

If values for k_s and k_s are hard to find these three simplifications are made:

1. $k_d = 0$. It is assumed that the demand is unaffected.
2. $k_s = -k_d$. Demand and supply equally affected.
3. $k_s = 0$. Supply unaffected.

In the case study by Ekvall and Andrae [9] it was assumed both for the primary Pb market and the Pb scrap market that $k_s = -k_d$. In practice, in the LCA tool Eco-Lab the activities shown in Tables 4.9 and 4.10 represented the Pb market and Pb scrap market, respectively. Table 4.9 shows that 50% of the Pb to battery manufacturing replaces Pb to Pb solder and that the remaining 50% contributes to an increase in total production of secondary and primary Pb.

Table 4.9 Primary Pb market as used in EcoLab [9]

Category	Direction	Substance	Quantity	Unit	Comment
Raw material	Input	Pb, primary	0.87	kg	
Raw material	Input	Pb, secondary	0.13	kg	
Product	Output	Pb, mixed pr. and sec. to production and use of Pb solder	−1	kg	
Product	Output	Pb, mixed pr. and sec to production and use of Pb back-up battery	2	kg	

Table 4.10 Pb scrap market as used in EcoLab [9]

Category	Direction	Substance	Quantity	Unit	Comment
Raw material	Input	Pb scrap excl. electr.	−1	kg	Scrap Pb from other sources than PBAs using Sn–Pb solder
Raw material	Input	Sn62 PBA	2	kg	Scrapped PBAs using Sn–Pb solder
Product	Output	Pb scrap	1	kg	Intended for the Pb scrap markets

Table 4.10 shows that 50% of the Pb from scrap PBAs replaces scrap Pb from other products, and that the remaining 50% contributes to an increase in total Pb recycling.

The data input needs for a CLCA are:

1. marginal consumers;
2. marginal producers, and
3. price elasticities of supply and demand for a good (in the example Pb and scrap Pb) that marginal consumers buy and marginal producers produce, k_s and k_d.

It is possible to generate a number of scenarios and vary the values of k_s and k_d. In this book the econometric topic is examined only superficially, and the reader will need to study the basic theories elsewhere. In 2007, Hübler and Frohn compiled a survey of developments within econometrics [15].

A very high k_d suggests that when the price of a good goes up, consumers will buy much less of it and when the price of that good goes down, consumers will buy much more. A very low k_d implies the opposite, that changes in price have little influence on demand:

- If $k_d > 1$ then demand is price elastic (demand *is* sensitive to price changes).
- If $k_d = 1$ then demand is unit elastic.
- If $k_d < 1$ then demand is price inelastic (demand *is not* sensitive to price changes).

k_s is used to see how sensitive the *supply* of a good is to a price change. The higher k_s, the more sensitive producers and sellers are to price changes. A very high k_s suggests that when the price of a good goes up, sellers will supply much less of the good and when the price of that good goes down, sellers will supply much more. A very low k_s implies the opposite, that changes in price have little influence on supply:

- If $k_s > 1$ then supply is price elastic (supply *is* sensitive to price changes).
- If $k_s = 1$ then supply is unit elastic.
- If $k_s < 1$ then supply is price inelastic (supply is not sensitive to price changes).

Long-term values of k_s and k_d can be estimated from supply-price and demand-price graphs for different commodities. Figures 4.9 and 4.10 show how this was done.

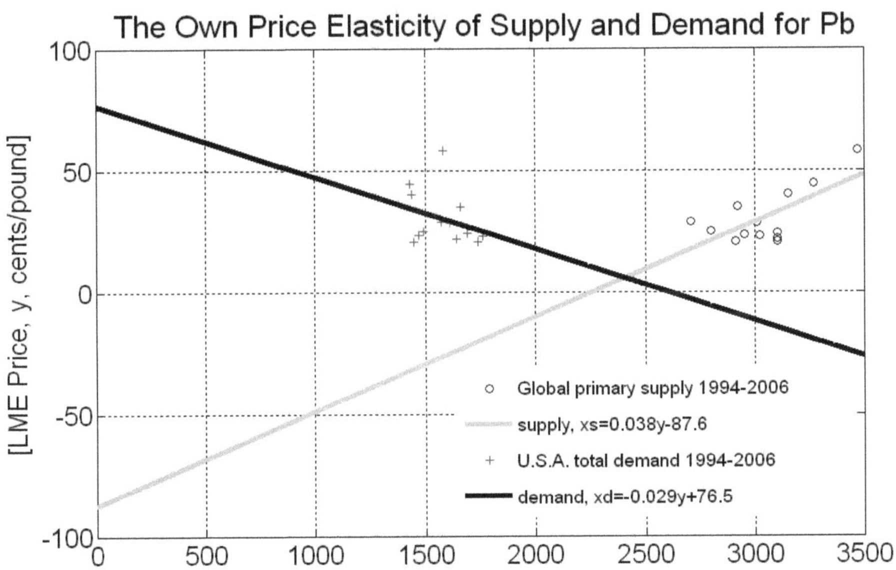

Fig. 4.9 Pb price as a function of the global primary Pb supply and the total US Pb demand

MATLAB® code for Fig. 4.9:

```
y = [24.8000   28.6000   35.1000   28.3000   24.0000
22.8000   20.6000   21.6000   20.5000   23.3000
40.2000   44.2   58];
```

```
xs = [2800          2710          2920          3010
3100          3020          3100          3100          2910
2950          3150   3270 3470];
xd =  [1490          1570          1660          1610
1690          1760          1740          1640          1450
1470          1440 1430 1580];
ps = polyfit(xs,y,1)
pd = polyfit(xd,y,1)
x2s = 0:0.5:3500;
x2d = 0:0.5:3500;
y2 = polyval(ps,x2s);
y3 = polyval(pd,x2d);
plot(xs,y,'o',x2s,y2,xd,y,'+',x2d,y3)
xlabel('Global Primary Supply, xs, and U.S.A.  Total
Demand, xd, ktonnes','FontSize',30)
ylabel('[LME Price, y, cents/pound]','FontSize',30)
title('The Own Price Elasticity of Supply and Demand
for Pb','FontSize',30)
legend('Global primary supply 1994-2006','supply,
xs=0.038y-87.6','U.S.A. total demand 1994-
2006','demand, xd=-0.029y+76.5')
grid on
```

Figure 4.9 suggests that both short-term k_d and k_s for Pb are less than 1 (–0.029 and 0.038, respectively). However, the data points are too few to draw any conclusions.

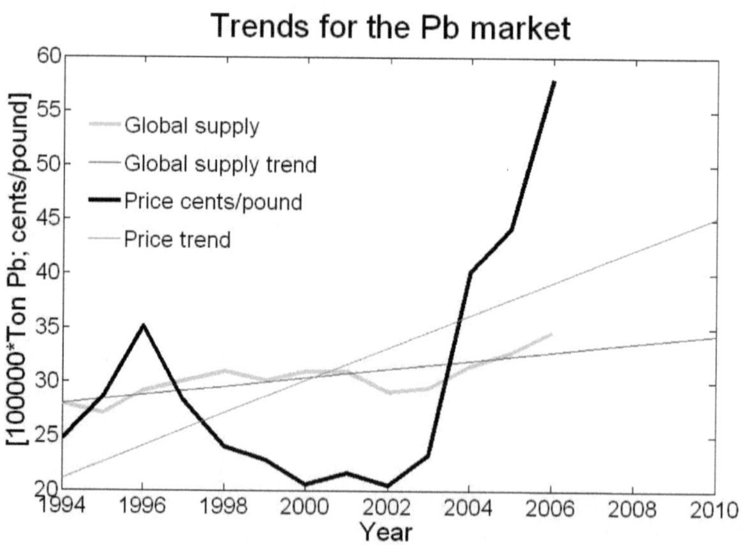

Fig. 4.10 Global supply of Pb and the Pb price as a function of year

MATLAB® code for Fig. 4.10:

```
x = [1994 1995 1996 1997 1998 1999 2000 2001 2002 2003
2004 2005 2006];%year
y1 = [28.00          27.10          29.20          30.10
31.00          30.20          31.00          31.00
29.10          29.50          31.50  32.70 34.70];%World Pb
supply
y11 = [24.8000   28.6000   35.1000   28.3000   24.0000
22.8000   20.6000   21.6000   20.5000   23.3000
40.2000   44.2   58];%Pb price cents per pound
p1 = polyfit(x,y1,1)
p6 = polyfit(x,y11,1)
x2 = 1994:2:2010;
y2 = polyval(p1,x2);
y12 = polyval(p6,x2);
plot(x,y1,'-',x2,y2, x,y11,'-',x2,y12)
xlabel('Year','FontSize',30)
ylabel('[100000*Ton Pb; cents/pound]','FontSize',30)
title('Trends for the Pb market','FontSize',30)
legend('Global supply','Global supply trend','Price
cents/pound','Price trend')
grid off
```

Figure 4.10 moreover suggests that the global Pb supply is insensitive to price changes (low k_s).

A description of the change in environmental burdens of the technosphere resulting from a shift from Sn–Pb solders to Pb-free solders requires a CLCI that includes an analysis of how the relevant markets are affected. Compared to ALCI the CLCI includes economic partial equilibrium models of the Pb and scrap Pb markets as well as the use of these goods outside the solder life cycle. For the partial equilibrium models, information about the sensitivity of the supply and demand, of Pb and scrap Pb, to price fluctuations is required. In order to include the alternative use of Pb outside the solder life cycle, information on how sensitive different Pb applications are to changes in the price of Pb is needed. A change in demand for a metal in the solder life cycle will affect the international markets at the margin. Thus, the initial data quality goal for many of the CLCI subsystems must be global marginal data.

In most cases, both supply and demand are more elastic long-term than short-term; the reason is that more factors can adapt to price changes in the long term: the production capacity of mines and power plants, the stock of household appliances, etc. High elasticity means that a customer can, if the price gets too high, change to something more cost efficient which gives the same function. A customer is sensitive to too high a price if the customer has a choice. On the other hand, if a supplier could not adapt to prices changes, the supplier would be inelastic. Long-term elasticity is in focus, as environmental systems analyses are primarily conducted due to concern about the long-term future environment.

4.4.1 Identification of Static Long-term Marginal Technologies

In order to describe how the environmental burdens of the technosphere are affected by a shift from Sn–Pb solders to Pb-free solders, the probable marginal production technologies for Pb were identified. Such technologies can be identified using, for example, a five-step procedure presented by Weidema et al. [16]. The procedure answers the overall questions:

1. What is the situation in which the studied change in demand occurs (steps a–c)?
2. Given this situation, what specific technology is actually affected by the change (steps d–e below)?

(a) What are the relevant time aspects? Example: either the long–term or the short–term marginal effects are practical to analyze.
(b) Specific processes or overall markets? For example, if a change only affects a single process, then the technology of that process is the marginal technology. If an entire market is affected then the marginal technology of that market must be identified. The market is identified by three steps [17]:

> *1. Define the external product in terms of obligatory properties*
> *2. Define market segments affected by the external product*
> *3. Define the competing product having identical obligatory properties as the external product*

(c) What is the trend in volume of the affected market?
(d) Which production has a potential to provide a capacity change?
(e) Which production is the most desirable?

4.4.2 Identification of Marginal Consumers

In addition, the marginal application of Pb outside the solder life cycles was to be identified. It is unclear whether a procedure exists for identifying the marginal use of a specific good. In [9] the marginal Pb consumers were identified by this procedure:

1. List the global uses for Pb (batteries, in glass, ammunition, piping, roofing, wall-cladding etc.).
2. List the alternatives, if there are any, for the global uses of Pb. The substitution effect is very important.
3. Estimate the share of Pb cost of the total production cost of the products.
4. Argue for which products the Pb price plays an important role.
5. Identify for which products the Pb price could remarkably affect the current and future competitiveness of the product.
6. Identify applications where there is a fierce competition with alternate technologies.

When using this procedure the alternative Pb use was identified as acid-Pb batteries for use as a power supply in combination with photovoltaic cells, in remote areas, stiffly competing with diesel generators, etc.

This is apparently too simple as the range of alternative Pb products also includes batteries for other applications, and probably also a significant proportion of roofing and wall cladding, where the use likely is particularly sensitive to a change in price.

As the marginal consumer(s) has been identified the so-called complementary product(s) and the competing product(s) must be identified.

The production, use and waste management of these products should also be included in the CLCA.

In the acid-Pb batteries case above, the complementary product(s) and the competing product(s) were photo-voltaic cells and electricity produced by diesel generators, respectively.

Sensitivity analyses can be carried out to test the importance of accurate k_s and k_d values to the importance of identifying the actual marginal processes.
Most important is to construct an econometric model of the costs and preferences of suppliers and buyers to be able to calculate the elasticities of supply and demand.

4.4.3 Sn Market

The recent trends in the global and US Sn markets are shown in Figs. 4.11 and 4.12. Figure 4.11 suggests that both short-term k_d and k_s for Sn are less than 1, however, the data points are too few to draw any conclusions. Sn demand in the US was more price sensitive than Pb supply between the years 1994 and 2006. The Sn price has been affected by the increased usage in the solder industry.

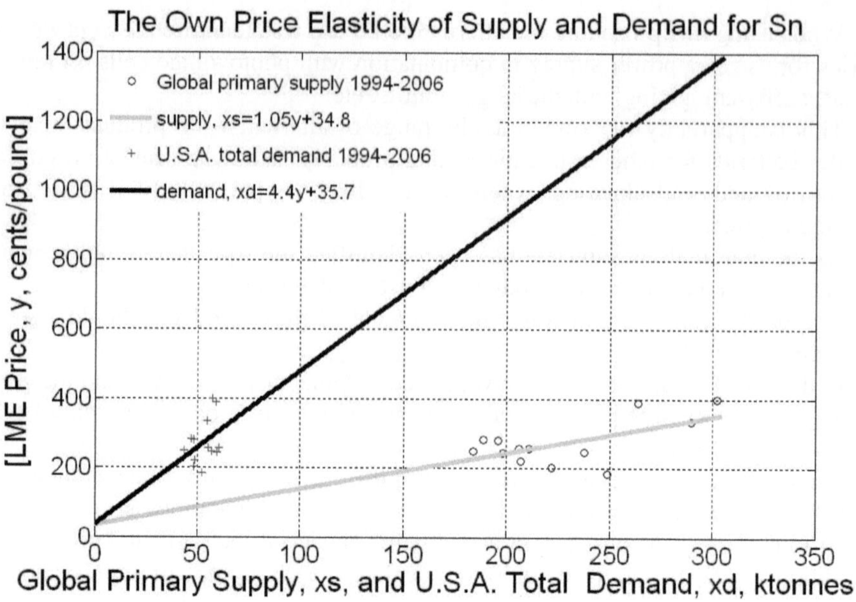

Fig. 4.11 Sn price as a function of global primary Sn supply and total US Sn demand

MATLAB® code for Fig. 4.11:

```
y = [248 282 279 256 256 245 246 203   184 222 388 334
398 ]; %Sn price, cents/pound
xs = [184  189 196  211 206 198  238  222 249  207 264
290 302];%global supply of Sn
xd =  [43.3  47   48.4 55.3 60.62 59.7  57.16 48.25
52.33  48.7 59.2 54.7 57.5];
ps = polyfit(xs,y,1)
pd = polyfit(xd,y,1)
x2s = 0:0.5:305;
x2d = 0:0.5:305;
y2 = polyval(ps,x2s);
y3 = polyval(pd,x2d);
plot(xs,y,'o',x2s,y2,xd,y,'+',x2d,y3)
xlabel('Global Primary Supply, xs, and U.S.A. Total
Demand, xd, ktonnes','FontSize',30)
ylabel('[LME Price, y, cents/pound]','FontSize',30)
title('The Own Price Elasticity of Supply and Demand
for Sn','FontSize',30)
legend('Global primary supply 1994-2006','supply,
xs=1.05y+34.8','U.S.A. total demand 1994-2006','demand,
xd=4.4y+35.7')
grid on
```

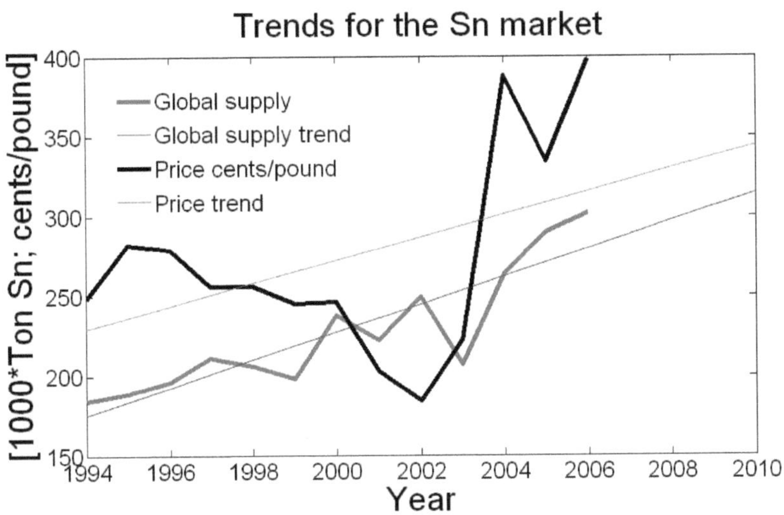

Fig. 4.12 The global supply of Sn and the Sn price as a function of year

MATLAB® code for Fig. 4.12:

```
x = [1994 1995 1996 1997 1998 1999 2000 2001 2002 2003
2004 2005 2006];%Year
y1 = [184  189 196  211 206 198  238  222 249  207 264
290 302];%global supply of Sn times 1000
y11 = [248  282 279 256 256 245 246 203   184 222 388
334 398 ]; %Sn price, cents/pound
p1 = polyfit(x,y1,1)
p6 = polyfit(x,y11,1)
x2 = 1994:2:2010;
y2 = polyval(p1,x2);
y12 = polyval(p6,x2);
plot(x,y1,'-',x2,y2, x,y11,'-',x2,y12)
xlabel('Year','FontSize',30)
ylabel('[1000*Ton Sn; cents/pound]','FontSize',30)
title('Trends for the Sn market','FontSize',30)
legend('Global  supply','Global  supply  trend','Price
cents/pound','Price trend')
grid off
```

4.4.4 Ag Market

As for annual Ag usage, industrial and decorative uses, photography, and jewellry and silver-ware represent more than 95%, and the electronics and photography industries are the main consumers [18]. The unique properties of Ag actually restrict its substitution in most applications.

The recent trends in the Ag markets are shown in Figs. 4.13 and 4.14. Figure 4.13 suggests that for Ag the short-term US k_d is more than 1, and k_s is less than 1. However, the data points are, as for Sn, too few to draw any conclusions. Ag demand is clearly more price sensitive than Pb demand.

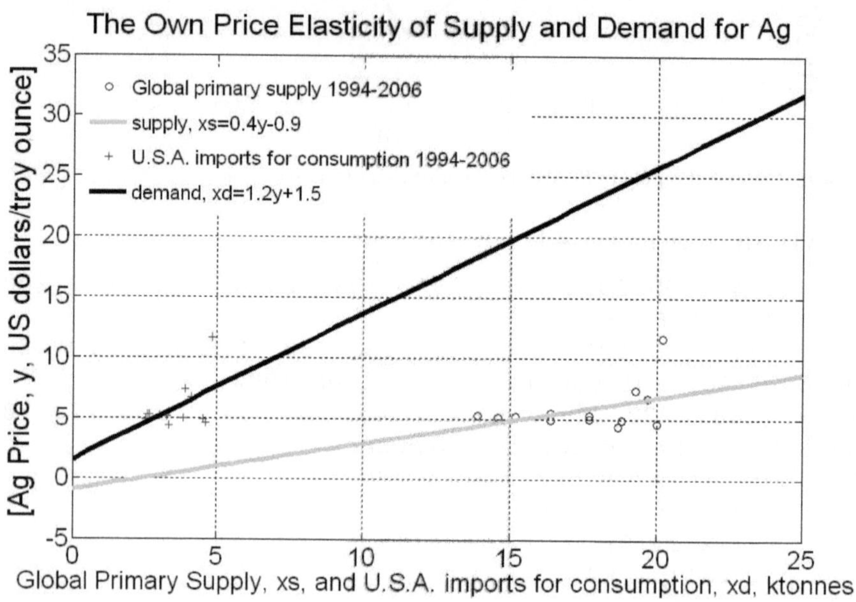

Fig. 4.13 Ag price as a function of global primary Sn supply and the total US Ag imports for consumption

MATLAB® code for Fig. 4.13:

```
y = [5.29 5.15 5.19 4.89 5.54 5.25 5.00 4.39    4.62
4.91 6.69 7.34 11.61 ]; %Ag price, US dollars/troy
ounce
xs = [13.9  14.6 15.2  16.4 16.4 17.7  17.7   18.7
20.0 18.8 19.7 19.3 20.2];%globally primary supply of
Ag, kton
xd = [2.6  3.25   3.01 2.54 3.33 2.66  3.81 3.31 4.60
4.51 4.10 3.88 4.82]; %US imports for consumption kton
ps = polyfit(xs,y,1)
pd = polyfit(xd,y,1)
x2s = 0:0.5:25;
```

```
x2d = 0:0.5:25;
y2 = polyval(ps,x2s);
y3 = polyval(pd,x2d);
plot(xs,y,'o',x2s,y2,xd,y,'+',x2d,y3)
xlabel('Global Primary Supply, xs, and U.S.A. imports
for consumption, xd, ktonnes','FontSize',30)
ylabel('[Ag Price, y, US dollars/troy
ounce]','FontSize',30)
title('The Own Price Elasticity of Supply and Demand
for Ag','FontSize',30)
legend('Global primary supply 1994-2006','supply,
xs=0.4y-0.9','U.S.A. imports for consumption 1994-
2006','demand, xd=1.2y+1.5')
grid on
```

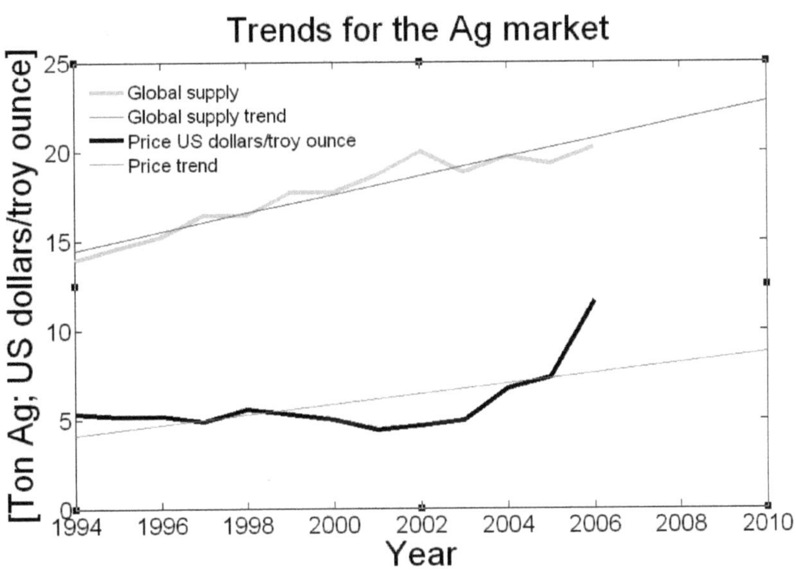

Fig. 4.14 Global supply of Ag and Ag price as a function of year

MATLAB® code for Fig. 4.14:

```
x = [1994 1995 1996 1997 1998 1999 2000 2001 2002 2003
2004 2005 2006];%Year
y1 = [13.9 14.6 15.2 16.4 16.4 17.7 17.7 18.7  20.0
18.8 19.7 19.3 20.2];%global primary supply of Ag, kton
y11 = [5.29  5.15 5.19 4.89 5.54 5.25 5.00 4.39   4.62
4.91 6.69 7.34 11.61 ]; %Ag price, US dollars/troy
ounce
p1 = polyfit(x,y1,1)
p6 = polyfit(x,y11,1)
x2 = 1994:2:2010;
```

```
y2 = polyval(p1,x2);
y12 = polyval(p6,x2);
plot(x,y1,'-',x2,y2, x,y11,'-',x2,y12)
xlabel('Year','FontSize',30)
ylabel('[Ton Ag; US dollars/troy ounce]','FontSize',30)
title('Trends for the Ag market','FontSize',30)
legend('Global supply','Global supply trend','Price US
dollars/troy ounce','Price trend')
grid off
```

Different market forces forced the Ag price to climb between the years 2005 and 2007. One of the explanations for the increase is the growth in Ag demand caused by the trend toward Pb-free solders and dominance of 3–4 wt% Ag solder alloys used by the electronics assembly industry.

The Ag cost increase of late raises the costs for solder materials suppliers. Sn and Ag represent the top two raw material costs for the solder industry.

4.4.5 Bi Market

In beginning of the 1990s in the United States and Europe, Bi was used to replace Pb everywhere where this was possible due to the latter's toxic properties. Figures 4.15 and 4.16 show that the global supply and US demand of Bi was price inelastic between 1994 and 2006. Naumov predicted that the Bi price will not increase if the Pb industry can provide the Bi demanded, otherwise significant price fluctuations will happen [19].

When 30–200 tonnes of Pb are produced around 1 tonne of Bi is obtained. This could imply that the amount of Bi available can not cover the intended Pb applications where Bi will be substituted. Pb from Bi mining will have to be disposed of and therefore would have limited commercial value, in turn raising the environmental and economic burdens of Bi.

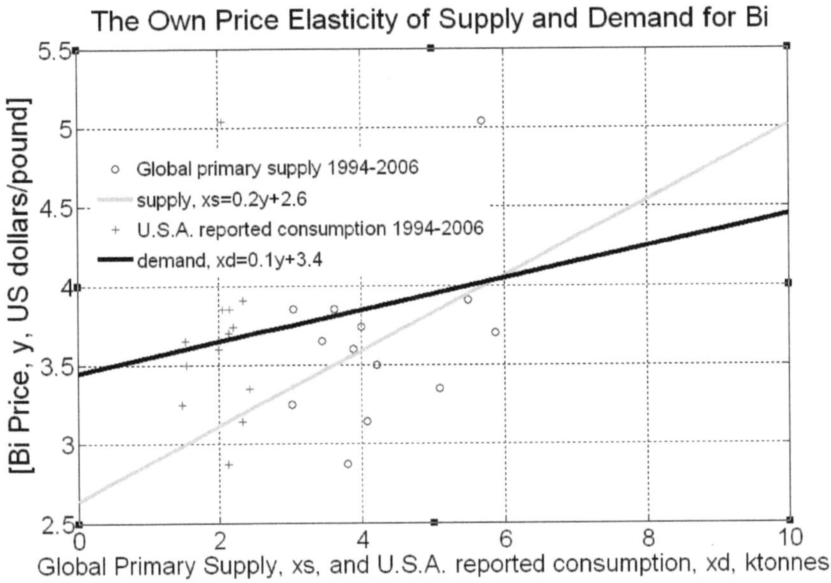

Fig. 4.15 Bi price as a function of global primary Bi supply and total US reported Bi consumption

MATLAB® code for Fig. 4.15:

```
y = [3.25   3.85 3.65 3.50 3.60 3.85 3.70 3.74  3.14
2.87 3.35 3.91 5.04 ]; %Bi price, US dollars/pound
xs = [3.02  3.04 3.44  4.21 3.88 3.62  5.88   4.00
4.07 3.80 5.10 5.50 5.70];%globally primary supply of
Bi, kton
xd =  [1.47  2.15 1.52 1.53 2.00 2.05  2.13 2.20 2.32
2.12 2.42 2.34 2.05]; %US imports for consumption kton
ps = polyfit(xs,y,1)
pd = polyfit(xd,y,1)
x2s = 0:0.5:10;
x2d = 0:0.5:10;
y2 = polyval(ps,x2s);
y3 = polyval(pd,x2d);
plot(xs,y,'o',x2s,y2,xd,y,'+',x2d,y3)
xlabel('Global Primary Supply, xs, and U.S.A. reported
consumption, xd, ktonnes','FontSize',24)
ylabel('[Bi Price, y, US dollars/pound]','FontSize',24)
title('The Own Price Elasticity of Supply and Demand
for Bi','FontSize',30)
legend('Global primary supply 1994-2006','supply,
xs=0.2y+2.6','U.S.A. reported consumption 1994-
2006','demand, xd=0.1y+3.4')
grid on
```

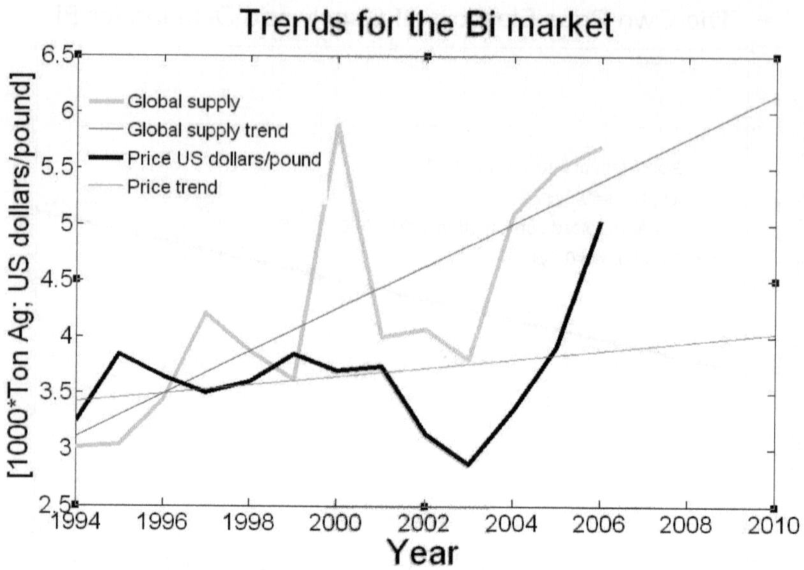

Fig. 4.16 Global supply of Bi and Bi price as a function of year

MATLAB® code for Fig. 4.16:

```
x = [1994 1995 1996 1997 1998 1999 2000 2001 2002 2003
2004 2005 2006];%Year
y1 = [3.02 3.04 3.44  4.21 3.88 3.62  5.88 4.00 4.07
3.80 5.10 5.50 5.70];%globally primary supply of Bi,
kton
y11 = [3.25 3.85 3.65 3.50 3.60 3.85 3.70 3.74 3.14
2.87 3.35 3.91 5.04 ]; %Bi price, US dollars/pound
p1 = polyfit(x,y1,1)
p6 = polyfit(x,y11,1)
x2 = 1994:2:2010;
y2 = polyval(p1,x2);
y12 = polyval(p6,x2);
plot(x,y1,'-',x2,y2, x,y11,'-',x2,y12)
xlabel('Year','FontSize',30)
ylabel('[1000*Ton Ag; US dollars/pound]','FontSize',30)
title('Trends for the Bi market','FontSize',30)
legend('Global supply','Global supply trend','Price US
dollars/pound','Price trend')
grid off
```

4.5 Input–Output Tables Expanded with Environmental Loadings

In 1973 Leontief received the Nobel Prize for his theory on the economy as sectors which are dependent on each other in input–output tables. LCA researchers have used Leontief's theory to create so-called environmentally expanded input–output tables (EEIO) [20].

Obviously EEIOLCI can give indications of indirect environmental loadings as all production is driven by a final consumption. Given the relatively few data and low resolution of current EEIO which is based on sector-level averages, it is difficult to separate two specific solder technologies.

EEIOLCI has nevertheless many advantages such as comprehensibility and rapid calculation, and the results obtained by the method should always be compared to the ones obtained by process-sum LCI.

It has been argued that EEIOLCI is more business oriented than process-sum LCI as it is based on the ready available accounting information which is ever present in the daily procedures in all organizations.

The mass and energy flows are thought to be less accessible. EEIOLCI and process-sum LCI can definitely complement each other. Some typical EEIOLCI loadings for interconnection materials are shown in Table 4.11 compared to process-sum ALCI and process-sum CLCI.

Table 4.11 The effect of different LCI models for the LCI result

LCA model	Interconnection material type	CO_2-equivalents [kg]/[kg]
Process-sum ALCA	62Sn–36Pb–2Ag	101
	95.5Sn–3.8Ag–0.7Cu	127
EEIOLCA (only cradle-to-gate for generic solder metal alloy)	62Sn–36Pb–2Ag	83
	95.5Sn–3.8Ag–0.7Cu	121
CLCA	62Sn–36Pb–2Ag	104
	95.5Sn–3.8Ag–0.7Cu	127

In Table 4.11 (for EEIOLCA) prices of \$80/kg and \$116/kg were used for 62Sn–36Pb–2Ag and 95.5Sn–3.8Ag–0.7Cu, respectively.

4.5.1 The Relation Between the EEIOLCI and the Consequential LCI

EEIOLCI is based on general equilibrium models which take into account the entire economy. CLCI is based upon/makes use of partial equilibrium models which study the price of one entity and assume that all other entity prices remain the same.

4.6 Eco-efficiency

LCA, environmental accounting, and eco-efficiency are all tools and methods representing environmental information on products and corporate programs in quantitative terms.

The concept of eco-efficiency can be defined in several ways. The philosophic foundation of the most well-known eco-efficiency concept was laid by the World Business Council for Sustainable Development in 1993, trying to provide a way of simultaneously estimating the financial profits and the environmental performance of industrial activities.

The simplest definition of eco-efficiency is the quotient of economic generation and ecological destruction [21].

Some argue that the life cycle costlife cycle (LCC) represents the value of the product (economic generation) and the environmental cost as calculated by LCA represents the ecological destruction. So far neither scientific consensus nor practical standards exist on how to estimate ecological destruction in monetary terms. Nevertheless, EPS and LIME (described in Sect. 3.3) are examples of two environmental damage-based attempts.

To exemplify, Germany as a nation decreased its CO_2 emissions between the years 2000 and 2007 and at the same time increased the net value between exports and imports. Germany arguably then increased its CO_2 eco-efficiency.

LCA is mainly used to assess the environmental aspects of products which use different technologies. However, the crucial point for firms is how to balance the economic and environmental aspects of products. Eco-efficiency is an attempt to indicate the relationship between the two aspects. Normally, eco-efficiency can be obtained by calculating the ratio of the value of a product or service targeted for assessment to the environmental impact.

Economic indicators such as the product price and its added value in addition to product functions like its cooling capacity (measured in terms of the floor area that can be cooled) are used as the numerator. The denominator of the eco-efficiency quotient is either represented by the results of characterization such as global warming, or the results of integration. However, many formulae for eco-efficiency actually have numerators and dominators whose dimensions are different from each other. Therefore, a dimensionless factor could be calculated as the ratio between the eco-efficiency of a product (new product) and the eco-efficiency of a standard product (old product). This ratio is usually defined as the eco-factor.

In Japan there are currently steady moves toward the utilization of eco-factors and eco-efficiency as indicators of corporate environmental management. In 2004, according to JEMAI, 67 companies had introduced eco-efficiency indicators and eco-factors on corporate or product level as an environmental management guide. Nevertheless, more companies publish environmental reports than utilize eco-efficiency indicators.

Moreover, the number of companies using eco-efficiency indicators has doubled from the previous year and is expected to grow further.

The most important issue is how to obtain the elements that constitute an indicator, either for eco-efficiency or an eco-factor. In some cases, eco-efficiency is calculated for an environment loading such as CO_2, or the amount of energy or, e.g., the consumption of resources.

However, this could increase the number of items in the assessment results and diminishes the lucidity of eco-efficiency as a communication tool. Therefore, several case studies in firms have been executed where the environmental impact is integrated by LIME and used as a denominator of eco-efficiency.

Companies in Japan have used LIME to calculate eco-efficiencies showing that they had reduced their environmental impact without losing economic profit. Companies can obtain so called eco-factors by calculating the eco-efficiency of various new products and comparing it with the eco-efficiency of old products.

4.7 Data Analysis in LCI – Dominance, Uncertainty, and Sensitivity

This section focuses on the practical approaches for dominance, uncertainty, and sensitivity analysis in LCA.

Several practical approaches exist for sensitivity and uncertainty analysis in life cycle inventory and impact assessment.

Table 4.12 shows that SimaPro, EMLCA, and CMLCA provide sensitivity values for different parts of the LCA. They all base their numerical values on perturbation analysis.

SimaPro does not provide sensitivity for all of environmental loadings, LCIA and weighting, as does CMLCA. EMLCA provides sensitivities of environmental loadings.

It was observed that most LCA software programs do not include the sensitivity step, but only dominance and often uncertainty analysis.

The practitioner has to use additional software tools to quantify the sensitivity priorities. Moreover, several attempts to formalize the LCI sensitive framework has been presented; however, it is not clear if generally dominance analysis give the same conclusions as sensitivity analysis [22].

Table 4.12 Sensitivity analysis coverage in LCA software

Software	Sensitivity for process	Sensitivity for environmental loadings	Sensitivity for characterization and weighting
SimaPro	Yes	No	No
CMLCA	Yes	Yes	Yes
EMLCA	Yes	Yes	No

4.7.1 Accuracy and Precision of Data in LCA

Figure 4.17 shows a suggestion for graphical representation of accuracy and precision.

In LCIA, accuracies do not exist as there are not yet known values which can be considered "true" for total systems. Accuracies exist nevertheless for measurements of thousands of individual data points. For example, it is somewhat unquantifiable what GWP100 score to expect when starting an LCA for a certain system.

There is no standard for which LCA results to expect from various products. On the other hand, precision is commonly used in LCA as it is a function of the standard deviation of measurement (data elements). Thus precision exist in the final LCIA/LCA results, however, so far accuracy does not.

4.7.2 Important Remarks About Software Capability of Calculating Uncertainty

To the author's knowledge, SimaPro 7 is the only LCA software which can calculate a correct uncertainty (probability) for a *comparison* between alternative "A" and "B."

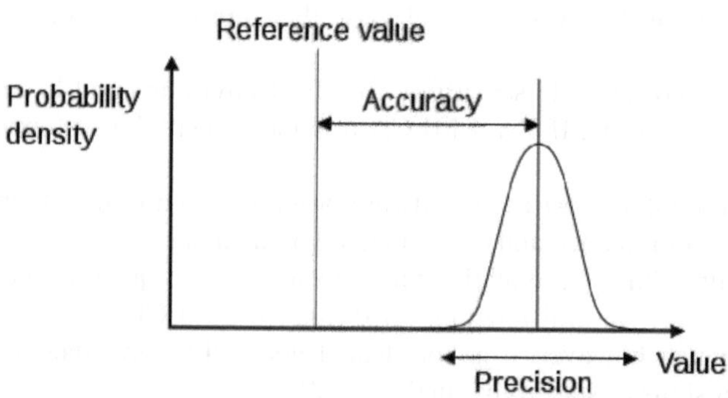

Fig. 4.17 Definitions of accuracy and precision

The reason is that SimaPro calculates the *difference between alternative "A" and "B" for each run* of the Monte Carlo simulation. That is so-called *process tree correlation* is used to ensure that the sampling variation is the same.

On the downside SimaPro does not support *process record correlation* in which the uncertainty of different consumptions and emissions in one unit process are modeled.

For many comparisons the user applies the same database (e.g., ecoinvent) to the product systems, e.g., identical electricity and transport modules. Then the data modules are correlated and dependent on each other.
The uncertainty would be overestimated if the probability of "A" being better/worse than "B" is calculated by comparing the absolute scores with uncertainty.

Absolute scores with uncertainty *can* be compared if the background systems are totally different, e.g., production of a component in different nations, electricity produced from wind, nuclear, etc.

EMLCA or CMLCA could be used; however then the uncertainty (for commonly used processes) for *one of the two* comparables has to be "eliminated." For example, if two solders are compared and they at some stage use an identical electricity production, for one of the solders the uncertainty factor must be eliminated before calculating the probability of one being better than the other. An example of this procedure is described in Sect. 4.1.4 for EMLCA and Pb emissions to water.

References

1. Haapala KR, Rivera JL, Sutherland JW (2008) Application of life cycle assessment tools to sustainable product design and manufacturing. Int J Innov Comput, Inform Control 4:575–589
2. Ciroth A (2007) ICT for environment in life cycle applications openLCA – a new open source software for life cycle assessment. Int J LCA 12:209–210
3. Andrae ASG (2005) Environmental life-cycle assessment in microelectronics packaging. PhD thesis, Chalmers University, Gothenburg, Sweden
4. Lu W (2006) Study on the advanced technique of environmental assessment based on life cycle assessment using matrix method. PhD thesis, The University of Tokyo, Tokyo, Japan
5. Sullivan CR (1994) Energy use: the most important impact of electronic equipment? Proc. IEEE Int Symp Electron Environ, 2–4 May, San Fransisco, CA, 261–266
6. Frischknecht R, Rebitzer G (2005) The ecoinvent database system: a comprehensive web-based LCA database. J Clean Prod 13:1337–1343
7. Wolf MA (2008) Supporting life cycle based policies and applications in industry with recommended LCI practice and the ELCD-consistent LCI data network. Pres at SETAC Europe Annual Meet, May 25–29, Warzaw, Poland
8. Huisman J, Stevels ALN, Stobbe I (2004) Eco-efficiency considerations on the end-of-life of consumer electronic products. IEEE Trans Electron Packag Manuf 27:9–25
9. Ekvall T, Andrae ASG (2006) Attributional and consequential environmental assessment of the shift to lead-free solders. Int J LCA 11:344–353
10. Lesage P, Deschenes L, Samson, R (2007) Evaluating holistic environmental consequences of brownfield management options using consequential life cycle assessment for different perspectives. Environmental Management 40:323–337
11. Sandén, BA, Karlstrom M (2007) Positive and negative feedback in consequential life-cycle assessment. J Clean Prod 15 :1469–1481
12. Eriksson O, Finnveden G, Ekvall T et al (2007) Life cycle assessment of fuels for district heating: A comparison of waste incineration, biomass- and natural gas combustion. Energy Policy 35:1346–1362

13. Thomassen MA, Dalgaard R, Heijungs R et al (2008) Attributional and consequential LCA of milk production. Int J LCA 13:339–349
14. Newcomer A, Blumsack SA, Apt J et al (2008) Short run effects of a price on carbon dioxide emissions from US electric generators. Environ Sci Technol 42:3139–3144
15. Hübler O, Frohn J (2007) Modern econometric analysis. Springer, Berlin, Heidelberg
16. Weidema BP, Frees N, Nielsen AM (1999) Marginal production technologies for life cycle inventories. Int J LCA 4:48–56
17. Ekvall T, Weidema BP (2004) System boundaries and input data in consequential life cycle inventory analysis. Int J LCA 9:161–171
18. Lanzano T, Bertram M, De Palo M et al (2006) The contemporary European silver cycle. Resour Conserv Recycl 46:27–43
19. Naumov A (2007) World market of bismuth: a review. Russ J Non-Ferr Met 48:10–16
20. Hendrickson CT, Lave LB, Matthews HS (2006) Environmental life cycle assessment of goods and services: an input–output approach. Resources for the Future Press
21. Park PJ, Tahara K (2008) Quantifying producer and consumer-based eco-efficiencies for the identification of key ecodesign issues. J Clean Prod 16:95–104
22. Heijungs R, Suh S (2002) The computational structure of life cycle assessment, Kluwer, Dordrecht

Chapter 5
LCA Case Studies of Solders

Abstract This chapter presents results of case studies which use the theories presented in Chaps. 3 and 4. Comparisons of Pb-based and Pb-free solders are done using CLCA, LIME, and air/water toxicity factors. The life cycle inventory calculations are shown in detail for chosen examples. Using the CMLCA software Ag-free solders are compared to the most common Sn–Ag–Cu combination as a starting point. The extensive interpretation capabilities of CMLCA are demonstrated, especially that uncertainties on all levels of LCA can be included. The key environmental performance indicators of solder pastes are suggested.

5.1 Pb Solder Compared to Pb-free Solder

As of July 1, 2006, Pb was banned in the majority of solders used in the electronics industry. This called for environmental evaluation of alternatives to Sn–Pb solders. The LCA conducted by Ekvall and Andrae [1] had two aims:

1. To compare a typical Sn–Pb solder paste of alloy composition 62wt% Sn, 36wt% Pb, 2wt% Ag to a Pb-free solder paste, 95.5 wt% Sn, 3.8 wt% Ag, 0.7 wt% Cu)
2. To compare attributional LCA (ALCA) and consequential (CLCA) methodologies

An ALCA model describes the environmental impact of the solder life cycle. Ideally, it should include average data on each unit process within the life cycle. A CLCA model aims at describing how the environmentally relevant physical flows to and from the technosphere will change in response to possible changes made within the solder life cycle. In Figs. 5.1 and 5.2 the system boundaries for the attributional and the consequential study are shown. The case study was performed in both EcoLab and EMLCA.

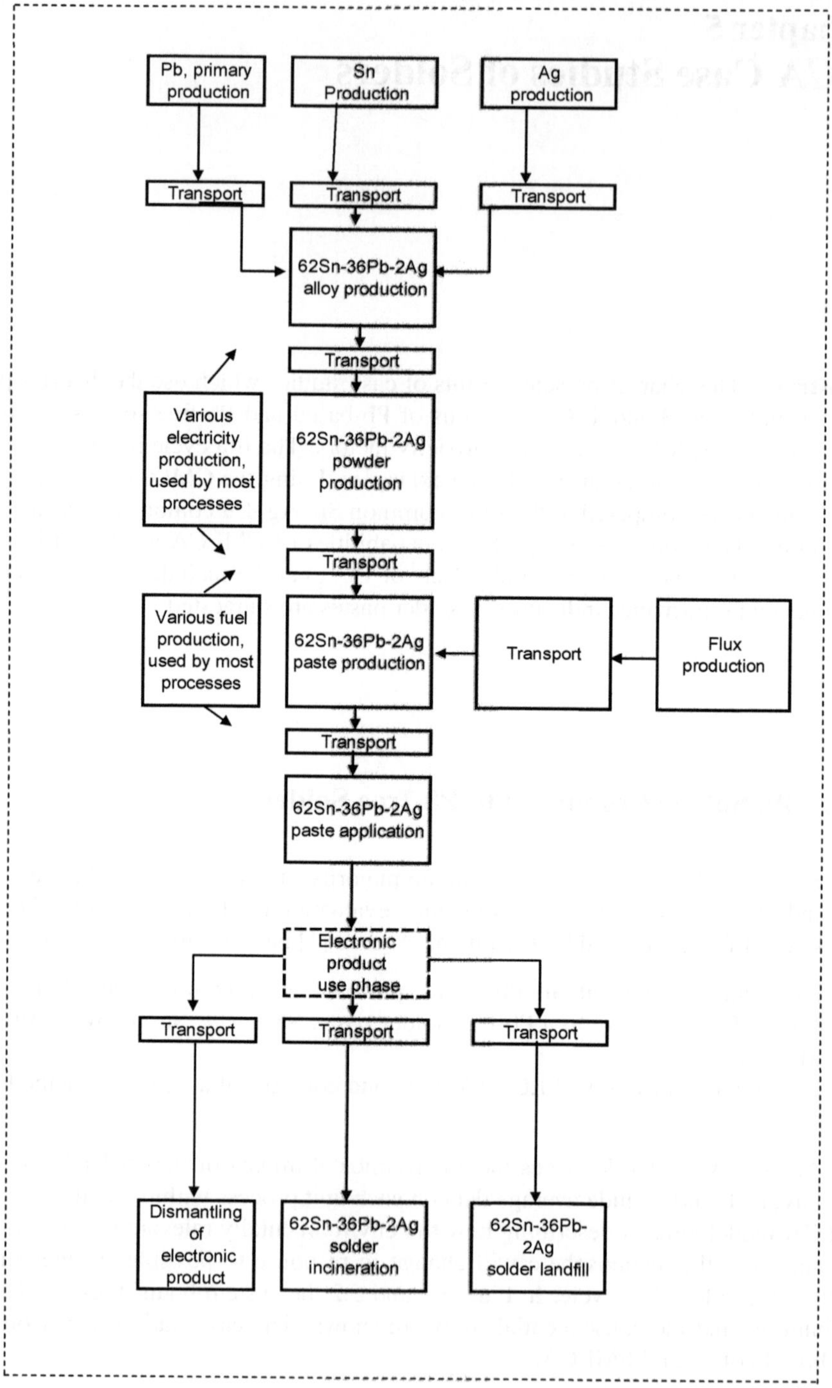

Fig. 5.1 System boundaries for the ALCA of 62Sn–36Pb–2Ag solder paste

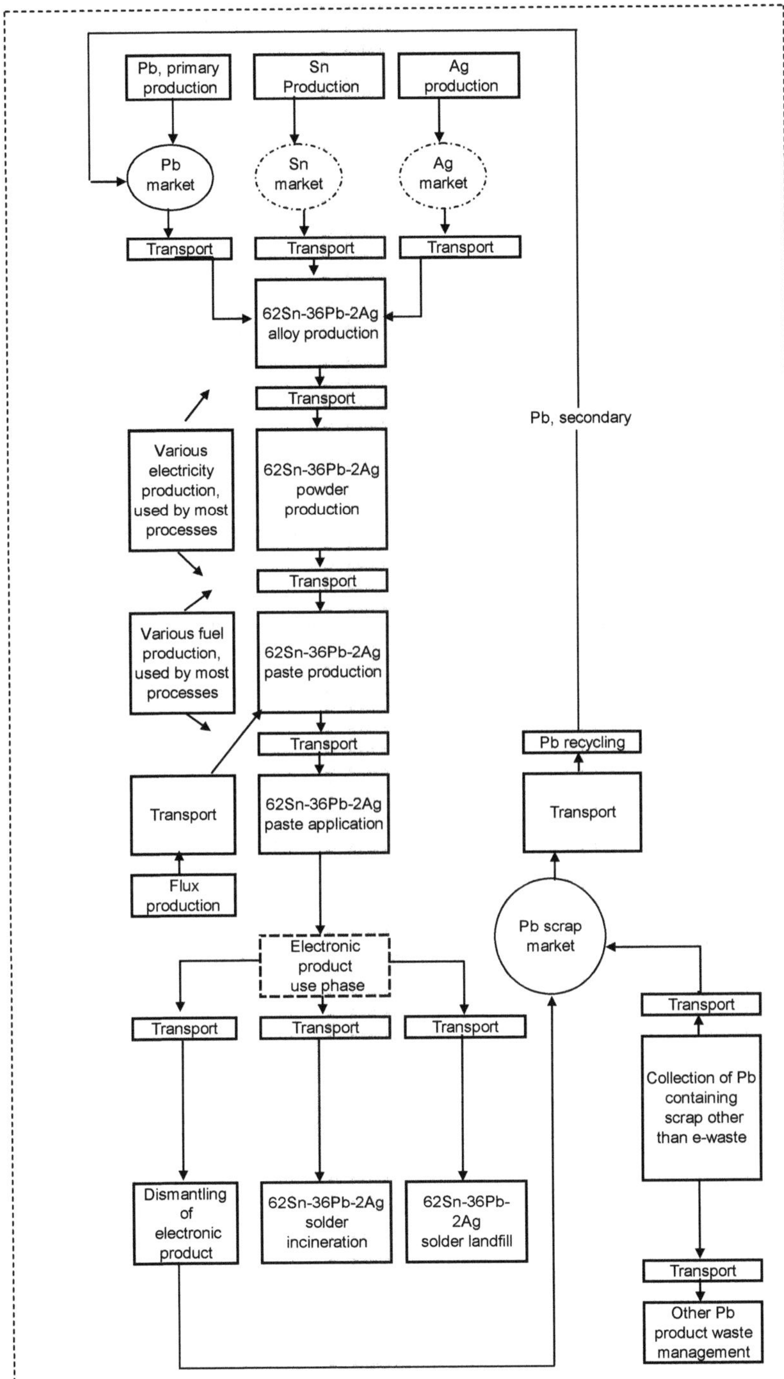

Fig. 5.2 System boundaries for the CLCA of 62Sn–36Pb–2Ag solder paste

The case study [1] did not eliminate any life cycle stages. However, it focused on Pb emissions and CO_2-equivalent emissions. Global warming potential was the only environmental impact category included and several sets of surrogate data were used.

5.1.1 Scope

The objects of the present case study used a cradle-to-grave system boundary. Global warming potential was the only impact assessment method explored due to lack of reliable data.

The model did not include unit processes other than those of the life cycle investigated, but significant cut-offs within the life cycle could be avoided through the use of environmentally expanded economic input–output tables for "flux production" [2].

A CLCA model includes unit processes that are significantly affected irrespective of whether they are within or outside the life cycle. Ideally, it should include marginal data on bulk production processes in the background system. The CLCA model included economic partial equilibrium models of the Pb and scrap Pb markets. However, both LCA models were based on data from the literature or from individual production sites. The partial equilibrium models were based on assumptions.

The functional unit was the volume (approximately 530 mm^3) of solder paste, corresponding to 2.5 grams of 62Sn–36Pb–2Ag solder paste needed to mount components onto a common printed wiring board.

Tables 5.1–5.3 show the LCI data, used in the case study, expressed in EMLCA inspired manner [1].

Table 5.1 Data used in ALCA model of 62Sn–36Pb–2Ag solder paste

PROCESS			
Electricity, global average			
F	Electricity, global average	MJ	1.00E+00
L	Carbon dioxide	kg	1.32E–01
L	Pb, air	kg	2.62E–08
L	Pb, soil	kg	2.39E–11
L	Pb, water	kg	6.22E–08
Natural gas production, global average			
F	Natural gas production, global average	MJ	1.00E+00
L	Carbon dioxide	kg	3.15E–03
L	Pb, air	kg	0.00E+00

Table 5.1 continued

PROCESS			
L	Pb, soil	kg	0.00E+00
L	Pb, water	kg	0.00E+00
Natural gas combustion, global average			
F	Natural gas combustion, global average	MJ	1.00E+00
I	Natural gas production, global average	MJ	–1.00E+00
L	Carbon dioxide	kg	5.91E–02
L	Pb, air	kg	0.00E+00
L	Pb, soil	kg	0.00E+00
L	Pb, water	kg	0.00E+00
Heavy fuel production, global average			
F	Heavy fuel production, global average	MJ	1.00E+00
L	Carbon dioxide	kg	1.50E–02
L	Pb, air	kg	5.23E–09
L	Pb, soil	kg	0.00E+00
L	Pb, water	kg	1.20E–08
Heavy fuel combustion, global average			
F	Heavy fuel combustion, global average	MJ	1.00E+00
I	Heavy fuel production, global average	MJ	–1.00E+00
L	Carbon dioxide	kg	7.79E–02
L	Pb, air	kg	1.00E–07
L	Pb, soil	kg	0.00E+00
L	Pb, water	kg	0.00E+00
Coal production, global average			
F	Coal production, global average	kg	1.00E+00
L	Carbon dioxide	kg	5.49E–02
L	Pb, air	kg	1.20E–08
L	Pb, soil	kg	0.00E+00
L	Pb, water	kg	0.00E+00

Table 5.1 continued

PROCESS			
Coal heat value			
F	Coal heat value	MJ	2.40E+01
I	Coal production, global average	kg	–1.00E+00
Coal combustion, global average			
F	Coal combustion, global average	MJ	1.00E+00
I	Coal heat value	MJ	–1.00E+00
L	Carbon dioxide	kg	9.15E+01
L	Pb, air	kg	3.00E–07
L	Pb, soil	kg	0.00E+00
L	Pb, water	kg	0.00E+00
production of Ag			
F	Ag	kg	1
L	Carbon dioxide	kg	10
L	Pb, air	kg	8.90E–04
L	Pb, water	kg	3.00E–05
L	Ag, in ground	kg	–1
production of Pb			
F	Pb	kg	1
I	Electricity, global average	MJ	–0.568
L	Carbon dioxide	kg	0.294
L	Pb, air	kg	2.54E–04
L	Pb, in ground	kg	–1
production of Sn			
F	Sn	kg	1
I	Electricity, global average	MJ	–75.6
L	Carbon dioxide	kg	4.6
L	Sn, in ground	kg	–1
production of 62Sn36Pb2Ag alloy			
F	62Sn36Pb2Ag alloy	kg	1
I	Electricity, global average	MJ	–1.075
I	Natural gas combustion, global average	MJ	–2.3485

Table 5.1 continued

PROCESS			
I	Pb	kg	–0.468
I	Sn	kg	–0.806
I	Ag	kg	–0.026
L	Pb, air	kg	1.52E–06
L	Carbon dioxide	kg	1.32E+00
L	Pb, water	kg	2.03E–07
production of 62Sn36Pb2Ag powder			
F	62Sn36Pb2Ag powder	kg	1
I	62Sn36Pb2Ag alloy	kg	–1.12
I	Natural gas combustion, global average	MJ	–0.3196
I	Electricity, global average	MJ	–3.1519
L	Pb, air	kg	1.25E–07
L	Carbon dioxide	kg	0.091
L	Pb, water	kg	1.57E–10
production of Flux for 62Sn36Pb2Ag powder			
L	Carbon dioxide	kg	3.80E+01
F	Flux for 62Sn36Pb2Ag powder	kg	1
62Sn36Pb2Ag solder paste production			
F	62Sn36Pb2Ag solder paste	kg	1.00E+00
I	Electricity, global average	MJ	–1.314
I	Natural gas combustion, global average	MJ	–5.67
I	Heavy fuel combustion, global average	MJ	–0.5399
I	62Sn36Pb2Ag powder	kg	–1.10E+00
I	Flux for 62Sn36Pb2Ag powder	kg	–1.01E–01
Stencil printing, 62Sn36Pb2Ag solder			
F	Stencil printing, 62Sn36Pb2Ag solder paste	kg	2.50E–03

Table 5.1 continued

PROCESS			
I	Electricity, global average	MJ	–5.76E–02
I	62Sn36Pb2Ag solder paste	kg	–2.83E–03
Reflow oven, 62Sn36Pb2Ag solder			
F	Reflow oven 62Sn36Pb2Ag solder	kg	2.25E–03
I	Stencil printing, 62Sn36Pb2Ag solder paste	kg	–2.50E–03
I	Electricity, global average	MJ	–1.20E+00
Usage of electronic product 62Sn36Pb2Ag solder			
F	Usage electronic product 62Sn36Pb2Ag solder	kg	1.00E+02
I	Reflow oven 62Sn36Pb2Ag solder	kg	–1.00E+02
O	Dismantling 62Sn36Pb2Ag solder	kg	1.00E+01
O	Incineration 62Sn36Pb2Ag solder	kg	1.40E+01
O	Landfill 62Sn36Pb2Ag solder	kg	7.60E+01
Landfill 62Sn36Pb2Ag solder			
F	Landfill 62Sn36Pb2Ag solder	kg	–1.00E+00
I	Electricity, global average	MJ	–2.87E–02
L	Pb, water	kg	6.77E–02
L	Sn, water	kg	1.80E–05
Incineration 62Sn36Pb2Ag solder			
F	Incineration 62Sn36Pb2Ag solder	kg	–1.00E+00
I	Electricity, global average	MJ	–3.89E–02
L	Pb, air	kg	9.90E–02
Dismantling 62Sn36Pb2Ag solder			
F	Dismantling 62Sn36Pb2Ag solder	kg	–1.00E+00
I	Electricity, global average	MJ	–6.70E–01
BOUNDARY			
B	Usage electronic product 62Sn36Pb2Ag solder	kg	2.25E–03

Table 5.2 continued

PROCESS			
Electricity, global average			
F	Electricity, global average	MJ	1
L	Carbon dioxide	kg	1.32E–01
L	Pb, air	kg	2.62E–08
L	Pb, soil	kg	2.39E–11
L	Pb, water	kg	6.22E–08
Natural gas production, global average			
F	Natural gas production, global average	MJ	1.00E+00
L	Carbon dioxide	kg	3.15E–03
L	Pb, air	kg	0
L	Pb, soil	kg	0
L	Pb, water	kg	0
Natural gas combustion, global average			
F	Natural gas combustion, global average	MJ	1
I	Natural gas production, global average	MJ	–1
L	Carbon dioxide	kg	5.91E–02
L	Pb, air	kg	0
L	Pb, soil	kg	0
L	Pb, water	kg	0
Heavy fuel production, global average			
F	Heavy fuel production, global average	MJ	1.00E+00
L	Carbon dioxide	kg	1.50E–02
L	Pb, air	kg	5.23E–09
L	Pb, soil	kg	0.00E+00
L	Pb, water	kg	1.20E–08
Heavy fuel combustion, global average			
F	Heavy fuel combustion, global average	MJ	1.00E+00
I	Heavy fuel production, global average	MJ	–1.00E+00
L	Carbon dioxide	kg	7.79E–02
L	Pb, air	kg	1.00E–07

Table 5.2 continued

PROCESS			
L	Pb, soil	kg	0.00E+00
L	Pb, water	kg	0.00E+00
Coal production, global average			
F	Coal production, global average	kg	1.00E+00
L	Carbon dioxide	kg	5.49E–02
L	Pb, air	kg	1.20E–08
L	Pb, soil	kg	0.00E+00
L	Pb, water	kg	0.00E+00
Coal heat value			
F	Coal heat value	MJ	2.40E+01
I	Coal production, global average	kg	–1.00E+00
Coal combustion, global average			
F	Coal combustion, global average	MJ	1.00E+00
I	Coal heat value	MJ	–1.00E+00
L	Carbon dioxide	kg	9.15E+01
L	Pb, air	kg	3.00E–07
L	Pb, soil	kg	0.00E+00
L	Pb, water	kg	0.00E+00
production of Ag			
F	Ag	kg	1
L	Carbon dioxide	kg	10
L	Pb, air	kg	8.90E–04
L	Pb, water	kg	3.00E–05
L	Ag, in ground	kg	–1
production of Sn			
F	Sn	kg	1
I	Electricity, global average	MJ	–75.6
L	Carbon dioxide	kg	4.6
L	Sn, in ground	kg	–1
L	Electricity, global average c	MJ	75.6
production of Cu			
F	Cu	kg	1.000
L	Cu, in ground	kg	–1.000

Table 5.2 continued

PROCESS			
I	Electricity, global average	MJ	–1.800
L	Carbon dioxide	kg	0.100
production of SAC alloy			
F	SAC alloy	kg	1
I	Electricity, global average	MJ	–1.075
I	Natural gas combustion, global average	MJ	–2.3485
I	Ag	kg	–0.0494
I	Sn	kg	–1.2145
I	Cu	kg	–0.0091
L	Pb, air	kg	1.52E–06
L	Carbon dioxide	kg	1.32E+00
L	Pb, water	kg	2.03E–07
production of SAC powder			
I	SAC alloy	kg	–1.12
F	SAC powder	kg	1
I	Electricity, global average	MJ	–3.1519
I	Natural gas combustion, global average	MJ	–0.3196
L	Pb, air	kg	1.25E–07
L	Carbon dioxide	kg	0.091
L	Pb, water	kg	1.57E–10
production of Flux for SAC powder			
L	Carbon dioxide	kg	3.80E+01
F	Flux for SAC powder	kg	1
SAC solder paste production			
I	Electricity, global average	MJ	–2.533
I	Natural gas combustion, global average	MJ	–5.39
I	Heavy fuel combustion, global average	MJ	–0.539
I	SAC powder	kg	–1.10E+00
I	Flux for SAC powder	kg	–1.10E–01
F	SAC solder paste	kg	1.00E+00

Table 5.2 continued

PROCESS			
Stencil printing, SAC solder			
I	Electricity, global average	MJ	−5.76E−02
I	SAC solder paste	kg	−2.49E−03
F	Stencil printing, SAC solder paste	kg	2.20E−03
Reflow oven, SAC solder			
I	Stencil printing, SAC solder paste	kg	−2.20E−03
I	Electricity, global average	MJ	−1.30E+00
F	Reflow oven SAC solder	kg	1.98E−03
Usage of electronic product SAC solder			
O	Dismantling SAC solder	kg	1.00E+01
O	Incineration SAC solder	kg	1.40E+01
O	Landfill SAC solder	kg	7.60E+01
I	Reflow oven SAC solder	kg	−1.00E+02
F	Usage electronic product SAC solder	kg	1.00E+02
Landfill SAC solder			
I	Electricity, global average	MJ	−2.87E−02
F	Landfill SAC solder	kg	−1.00E+00
L	Ag, water	kg	7.00E−07
L	Sn, water	kg	2.79E−05
L	Cu, water	kg	1.00E−07
Incineration SAC solder			
F	Incineration SAC solder	kg	−1.00E+00
I	Electricity, global average	MJ	−3.89E−02
L	Ag, air	kg	1.05E−02
L	Sn, air	kg	2.63E−01
L	Cu, air	kg	1.93E−03
Dismantling SAC solder			
I	Electricity, global average	MJ	−6.70E−01
F	Dismantling SAC solder	kg	−1.00E+00
BOUNDARY			
B	Usage electronic product SAC solder	kg	1.98E−03

Table 5.3 Data used in the CLCA model of 62Sn–36Pb–2Ag solder paste

PROCESS			
Electricity, global average			
F	Electricity, global average	MJ	1.00E+00
L	Carbon dioxide	kg	1.32E–01
L	Pb, air	kg	2.62E–08
L	Pb, soil	kg	2.39E–11
L	Pb, water	kg	6.22E–08
Natural gas production, global average			
F	Natural gas production, global average	MJ	1.00E+00
L	Carbon dioxide	kg	3.15E–03
L	Pb, air	kg	0.00E+00
L	Pb, soil	kg	0.00E+00
L	Pb, water	kg	0.00E+00
Natural gas combustion, global average			
F	Natural gas combustion, global average	MJ	1.00E+00
I	Natural gas production, global average	MJ	–1.00E+00
L	Carbon dioxide	kg	5.91E–02
L	Pb, air	kg	0.00E+00
L	Pb, soil	kg	0.00E+00
L	Pb, water	kg	0.00E+00
Heavy fuel production, global average			
F	Heavy fuel production, global average	MJ	1.00E+00
L	Carbon dioxide	kg	1.50E–02
L	Pb, air	kg	5.23E–09
L	Pb, soil	kg	0.00E+00
L	Pb, water	kg	1.20E–08
Heavy fuel combustion, global average			
F	Heavy fuel combustion, global average	MJ	1.00E+00
I	Heavy fuel production, global average	MJ	–1.00E+00
L	Carbon dioxide	kg	7.79E–02

Table 5.3 continued

PROCESS			
L	Pb, air	kg	1.00E–07
L	Pb, soil	kg	0.00E+00
L	Pb, water	kg	0.00E+00
Coal production, global average			
F	Coal production, global average	kg	1.00E+00
L	Carbon dioxide	kg	5.49E–02
L	Pb, air	kg	1.20E–08
L	Pb, soil	kg	0.00E+00
L	Pb, water	kg	0.00E+00
Coal heat value			
F	Coal heat value	MJ	2.40E+01
I	Coal production, global average	kg	–1.00E+00
Coal combustion, global average			
F	Coal combustion, global average	MJ	1.00E+00
I	Coal heat value	MJ	–1.00E+00
L	Carbon dioxide	kg	9.15E+01
L	Pb, air	kg	3.00E–07
L	Pb, soil	kg	0.00E+00
L	Pb, water	kg	0.00E+00
production of Ag			
F	Ag	kg	1
L	Carbon dioxide	kg	10
L	Pb, air	kg	8.90E–04
L	Pb, water	kg	3.00E–05
L	Ag, in ground	kg	–1
production of Pb 1st			
F	Pb 1st	kg	1.00E+00
L	Carbon dioxide	kg	4.95E–01
L	Pb, air	kg	2.85E–04
L	Pb, water	kg	1.80E–02
L	Pb, in ground	kg	1.00E+00
production of Sn			
F	Sn	kg	1.00E+00
L	Carbon dioxide	kg	1.46E+01
L	Sn, in ground	kg	1.00E+00

Table 5.3 continued

PROCESS			
L	Pb, air	kg	2.02E–05
L	Pb, water	kg	5.70E–07
production of 62Sn36Pb2Ag alloy			
F	62Sn36Pb2Ag alloy	kg	1.00E+00
I	Electricity, global average	MJ	–1.08E+00
I	Pb, mixed 1st and 2nd	kg	–0.468
I	Natural gas combustion, global average	MJ	–2.3485
I	Sn	kg	–0.806
I	Ag	kg	–0.026
L	Pb, air	kg	1.52E–06
L	Carbon dioxide	kg	1.32E+00
L	Pb, water	kg	2.03E–07
production of 62Sn36Pb2Ag powder			
F	62Sn36Pb2Ag powder	kg	1.00E+00
I	62Sn36Pb2Ag alloy	kg	–1.12E+00
I	Electricity, global average	MJ	–3.15E+00
I	Natural gas combustion, global average	MJ	–0.3196
L	Pb, air	kg	1.25E–07
L	Carbon dioxide	kg	9.10E–02
L	Pb, water	kg	1.57E–10
production of Flux for 62Sn36Pb2Ag powder			
F	Flux for 62Sn36Pb2Ag powder	kg	1.00E+00
L	Carbon dioxide	kg	3.80E+01
62Sn36Pb2Ag solder paste production			
F	62Sn36Pb2Ag solder paste	kg	1.00E+00
I	Electricity, global average	MJ	–1.31E+00
I	Natural gas combustion, global average	MJ	–5.67
I	Heavy fuel combustion, global average	MJ	–0.5399
I	62Sn36Pb2Ag powder	kg	–1.10E+00
I	Flux for 62Sn36Pb2Ag powder	kg	–1.01E–01

Table 5.3 continued

PROCESS			
Stencil printing, 62Sn36Pb2Ag solder			
F	Stencil printing, 62Sn36Pb2Ag solder paste	kg	2.50E–03
I	Electricity, global average	MJ	–5.76E–02
I	62Sn36Pb2Ag solder paste	kg	–2.83E–03
L	Electricity, global average c	MJ	5.76E–02
Reflow oven, 62Sn36Pb2Ag solder			
F	Reflow oven 62Sn36Pb2Ag solder	kg	2.25E–03
I	Stencil printing, 62Sn36Pb2Ag solder paste	kg	–2.50E–03
I	Electricity, global average	MJ	–1.20E+00
Usage of electronic product 62Sn36Pb2Ag solder			
F	Usage electronic product 62Sn36Pb2Ag solder	kg	1.00E+02
I	Reflow oven 62Sn36Pb2Ag solder	kg	–1.00E+02
O	Collection PBAs 62Sn36Pb2Ag solder	kg	1.00E+01
O	Incineration 62Sn36Pb2Ag solder	kg	1.40E+01
O	Landfill 62Sn36Pb2Ag solder	kg	7.60E+01
Landfill 62Sn36Pb2Ag solder			
F	Landfill 62Sn36Pb2Ag solder	kg	–1.00E+00
I	Electricity, global average	MJ	–2.87E–02
L	Pb, water	kg	6.77E–02
L	Sn, water	kg	1.80E–05
Incineration 62Sn36Pb2Ag solder			
F	Incineration 62Sn36Pb2Ag solder	kg	–1.00E+00
I	Electricity, global average	MJ	–3.89E–02
L	Pb, air	kg	9.90E–02
L	Sn, air	kg	1.71E–01
Collection 62Sn36Pb2Ag solder			
F	Collection PBAs 62Sn36Pb2Ag solder	kg	–1.00E+00
I	Electricity, global average	MJ	–6.70E–01
O	PBAs with 62Sn36Pb2Ag solder	kg	1.00E+00
PBAs with 62Sn36Pb2Ag solder			

Table 5.3 continued

PROCESS			
F	PBAs with 62Sn36Pb2Ag solder	kg	–1.00E+00
Scrap collection of other Pb cont. waste than electronics			
F	Scrap other than electronics cont. Pb	kg	7.50E–01
O	Scrap other than electronics cont. Pb to waste	kg	2.50E–01
Marginal waste treatment of Pb waste other than electronics			
F	Scrap other than electronics cont. Pb to waste	kg	–1.78E+07
I	Electricity, global average	MJ	–5.11E+05
L	Pb, water	kg	6.00E–02
Pb scrap market			
F	Pb scrap	kg	1.00E+00
I	PBAs with 62Sn36Pb2Ag solder	kg	–2.00E+00
I	Scrap other than electronics cont. Pb	kg	–1.00E+00
Pb recycling, production of Pb 2nd			
F	Pb 2nd	kg	1.00E+00
I	Pb scrap	kg	–1.08E+00
I	Electricity, global average	MJ	–1.08E+00
L	Carbon dioxide	kg	1.32E+00
L	Pb, water	kg	2.03E–07
Pb market, mixed 1st 2nd			
F	Pb, mixed 1st and 2nd	kg	1.00E+00
I	Pb 2nd	kg	–1.70E–01
I	Pb 1st	kg	–8.30E–01
production ABS/PC			
F	ABS/PC	kg	1.00E+00
L	Carbon dioxide	kg	4.91E+00
L	Pb, air	kg	3.80E–06
L	Pb, water	kg	2.50E–07
production of H2SO4			
F	H2SO4	kg	1.00E+00
L	Carbon dioxide	kg	3.00E–02
production of Pb acid battery			

Table 5.3 continued

PROCESS			
F	Pb acid battery	2V cells	8.50E+05
I	Pb, mixed 1st and 2nd	kg	–4.36E+06
I	Electricity, global average	MJ	–2.50E+06
I	H2SO4	kg	–9.82E+05
I	ABS/PC	kg	–4.01E+05
L	Carbon dioxide	kg	0.00E+00
L	Pb, air	kg	1.22E+03
L	Pb, water	kg	1.05E+02
Use of Solar cell and battery to generate electricity			
F	Use of Solar cell and battery to generate electricity	MJ	4.64E+03
I	Pb acid battery	2V cells	–2.86E+01
I	Solar cells, 49Wp	m2	–4.80E–01
I	Production of electricity from diesel combustion	MJ	–4.64E+03
Production solar photovoltaic cells			
F	Solar cells, 49Wp	m2	1.00E+03
L	Carbon dioxide	kg	2.93E+02
Production of electricity from diesel combustion			
F	Production of electricity from diesel combustion	MJ	–1.00E+00
L	Carbon dioxide	kg	3.33E–01
BOUNDARY			
B	Usage electronic product 62Sn36Pb2Ag solder	kg	2.25E–03
	Use of Solar cell and battery to generate electricity	MJ	2.55E–02

Here follows an explanation on the calculation of energy per Pb content, i.e., the value near "BOUNDARY" in Table 5.3 for "Use of Solar cell and battery to generate electricity." According to an anonymous battery manufacturer 4,374,666.5 kg Pb is used per 850,000 2V cells. A Pb-acid battery (100 Ah type) weighed 30 kg corresponding to 30/7 pieces of 2V cells as a 2V cell weighs 7 kg.

In the solar cell life cycle, 20/3 Pb-acid batteries are used during 20 years, i.e., 20/3×30/7 2V cells = 28.57 2V cells. That is 28.57 2V cells are needed per 1,288 kWh output electricity during 20 years. The Pb per functional unit (2.5 g Sn–Pb) is 0.9×2.5×0.36 = 0.81 g. 5.15 kg Pb is used per 2V cell, i.e., 147 kg Pb per 28.57 2V cells and per 1,288 kWh output electricity during 20 years. 0.81 g Pb then represents ((1,288 kWh×3.6 MJ/kWh)/147 kg Pb)×0.81 g Pb/F.U. = 0.0255 MJ/F.U.

5.1.2 Results and Discussion

The ALCA demonstrated the apparent fact that the shift from Sn–Pb to Pb-free meant that Pb was more or less eliminated from the solder life cycle. The ALCA results also indicated that Pb-free contributed 10% more to the global warming potential than Sn–Pb. Figure 5.3 shows the CO_2 emissions for the pastes in Tables 5.1 and 5.2, respectively. That confirmed earlier ALCAs comparing Sn–Pb and Pb-free solder pastes [3].

Despite the poor quality of the econometric and marginal LCI data, the CLCA demonstrated that, when Pb use is eliminated from the solder life cycle, the effect was partly offset by increased Pb use in batteries and other products.

Figure 5.4 shows the consequential CO_2 emission scores resulting from Tables 5.2 and 5.3.

Nevertheless, this Pb-free shift could contribute to environmental improvement because Pb emissions are likely to be greatly reduced. Moreover, batteries can contribute to reducing global warming potential, thereby offsetting part of the global warming potential increase in the solder life cycle.

5.1.3 Conclusions

ALCA and CLCA yielded complementary knowledge on the consequences of this shift in solder pastes. The shift from Sn–Pb solder paste to Pb-free is likely to result in reduced Pb emissions and increased global warming potential.

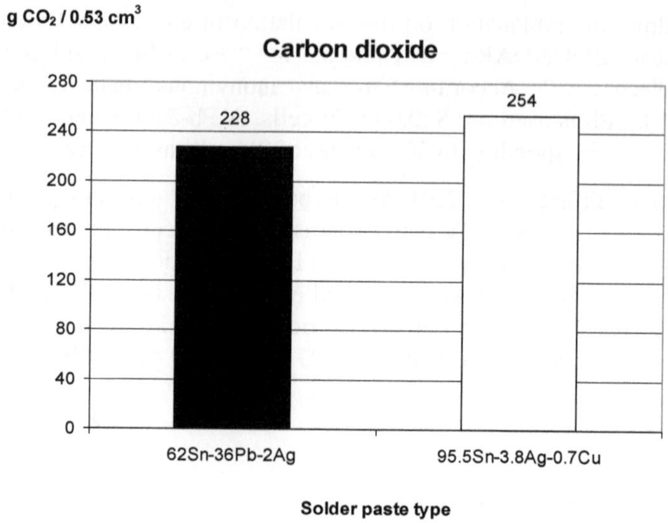

Fig. 5.3 Carbon dioxide emissions from solder paste life cycles

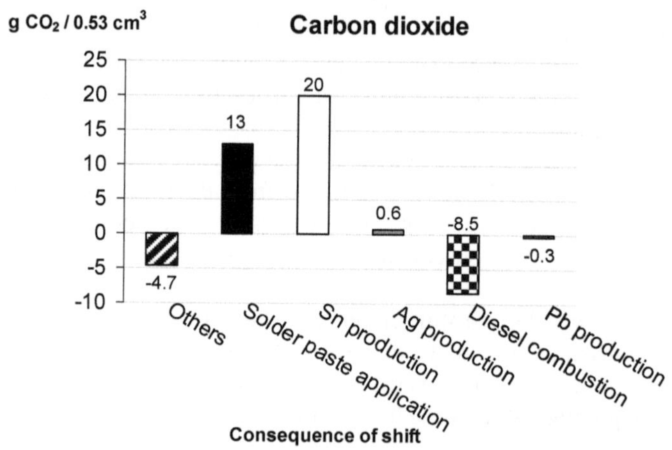

Fig. 5.4 Consequential CO_2 results when subtracting CLCA for 95.5Sn–3.8Ag–0.7Cu from CLCA of 62Sn–36Pb–2Ag

At present, CLCA is hampered by the lack of readily available marginal data and the lack of input data to economic partial equilibrium models. However, when the input to a CLCA model is in the form of quantitative assumptions based on a semi-qualitative discussion, the model can still generate new knowledge.

5.2 LIME Evaluation of the Shift to Pb-free

The hypothesis for the present research [4] was that there would be negative environmental impact assessment results globally, as a consequence of introducing 95.5Sn–3.8Ag–0.7Cu paste to replace 63Sn–37Pb paste.

5.2.1 Scope

The scope for the present case study was the same as for [1] shown in Sect. 5.1. The study was performed in EcoLab.

5.2.2 Results and Discussion

As shown in Figs. 5.5 and 5.6, the ALCA and the CLCA results both indicate that the overall LIME scores in JPY will increase considerably, almost 90%, per functional unit. The most important differences between 63Sn–37Pb and 95.5Sn–3.8Ag–0.7Cu are two-fold: Sn and Ag resource consumption. The Ag resources from Ag production are significant due to the relatively high integrated LIME factor for Ag, about 6,900 JPY/kg (Tables 3.2 and 3.3). CO_2 emissions from electricity production are also noticeable.

However, considering global warming results alone, it was earlier reported for a similar inventory to the present study that the solder application process and the Sn production were the processes mainly affected [1]. The marginal Pb usage, as a result of the solder Pb ban, will be where the competition is the largest and where the Pb consumers are most sensitive to a Pb price change.

On the margin, in remote areas, diesel combustion to generate electricity is supposed to be replaced by Pb back-up batteries (the marginal Pb usage) when Pb is banned in solders. All impacts attributable to Pb production will not disappear as on the margin Pb will be used in Pb-acid batteries (instead of solder pastes) which in combination with photovoltaic cells will replace the diesel combustion electricity. In fact, CLCA helped identify this offset in the impact related to the Pb production.

The present CLCA study predicts that the LIME score will rise as the avoidance of the environmental impacts connected to diesel combustion does not outweigh the increased resource consumptions and emissions from Ag, Sn, and energy production.

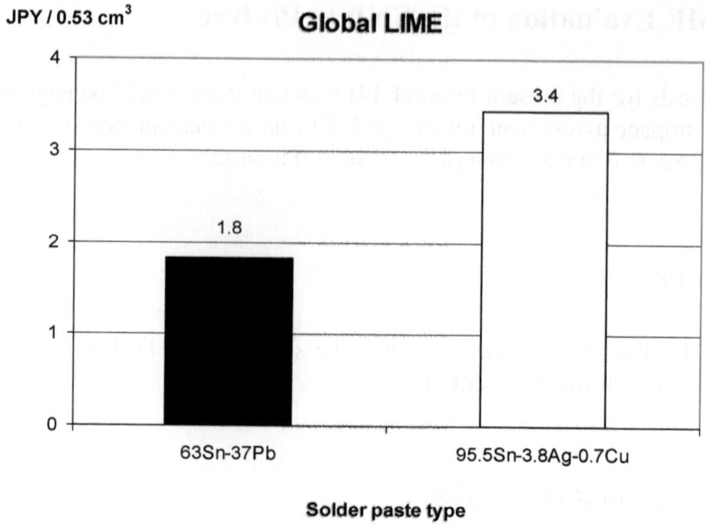

Fig. 5.5 Attributional LIME results for 63Sn–37Pb and 95.5Sn–3.8Ag–0.7Cu

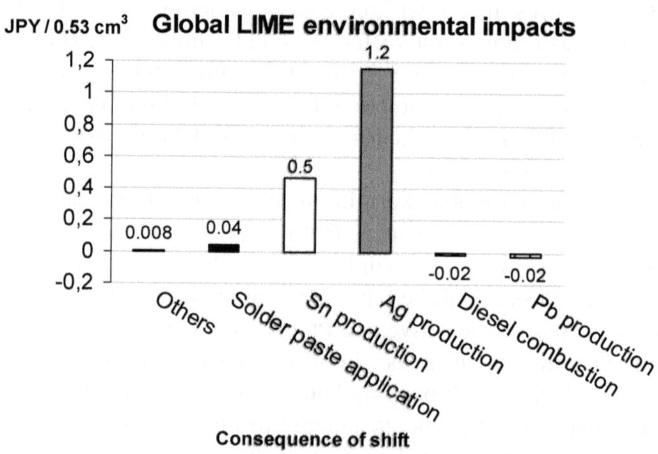

Fig. 5.6 Consequential LIME results when subtracting CLCA of 63Sn–37Pb from CLCA of 95.5Sn–3.8Ag–0.7Cu

Moreover, electricity extraction could be of higher importance, although not likely to change any conclusions. The data quality is poor for the model of the scrap Pb market and waste management of competing sources of scrap Pb. However, this part of the model does not have a big influence on the results.

5.2.2.3 Effect of Recycling

It is uncertain how much of the Sn, Ag, Pb, and Cu solder metals that will be recycled will specifically be used for production of new solders. Quantification of this is especially important for Ag and Sn. Based on this research increased recycling of Ag and Sn could be relevant to decreasing the global impact. In the present research it was not possible to apply the LIME factors for other than global effects, as the local LIME is adjusted for Japanese conditions. This screening of the global situation was, however, useful as it strengthens earlier results showing that the social and economic impacts, due to the consumption of resources for Sn and Ag production, could rise as a result of the shift to Pb-free solder paste.

Comparing the ALCA results for 63Sn–37Pb and 95.5Sn–3.8Ag–0.7Cu is not equal to predicting the consequences of shifting from 63Sn–37Pb to 95.5Sn–3.8Ag–0.7Cu. Occasionally, as seems to be the case regarding the solders in the present study, the two techniques, ALCA and CLCA, more or less provide the same conclusion. It is obvious that ALCA is not capable of estimating future environmental impacts, but on the other hand CLCA needs to be more developed in co-operation with econometricians to be more precise and comprehensive.

5.2.3 Conclusions

As far as globally related environmental impacts are concerned, the shift from 63Sn–37Pb to 95.5Sn–3.8Ag–0.7Cu solder paste is likely to increase them. A significant increase was detected due to increased Ag and Sn production.

No significant increase in the LIME score could be related to the increased generation of electricity for the 95.5Sn–3.8Ag–0.7Cu solder application processes as compared to 63Sn–37Pb. The LIME score is highly dependent on the Ag and Sn weighting factors. Ozone-depleting substances had an insignificant influence on the LIME score. This study confirms earlier work [3] reporting that the resource consumption will be higher for Sn–Ag–Cu solder pastes than conventional Sn–Pb. Comparing the possible effect of local Pb emissions to the global LIME impacts, the share of Pb emissions reaching the water environment from landfills in Tables 5.1 and 5.3 has to be 100 times smaller for the LIME score in shown in Fig. 5.6 not to be affected considerably. That is, should the Pb water emissions in Tables 5.1 and 5.3 be multiplied with LIME factors from Tables 3.2 and 3.3, the *LIME* scores of Fig. 5.6 would be totally dominated by these emissions.

5.3 Toxicity Evaluations

The toxicity indicators "OSHA Air Toxicity" (AT) [5] and "WHO Water Toxicity" (WT) [5] were applied to the solder pastes 62Sn–36Pb–2Ag and 95.5Sn–3.8Ag–0.5Cu. The hypothesis was that the scores would decrease, as a consequence of introducing 95.5Sn–3.8Ag–0.7Cu paste to replace 62Sn–36Pb–2Ag. The study was performed in EcoLab.

5.3.1 Scope

Probably most metals, including the one used for solders, are part of the global metal market and can end up in other applications as well.

The toxicity indicators AT and WT were estimated for ALCA and CLCA of solder pastes. By finding limit values for maximum allowable air concentrations. Due to lack of economic and marginal data, e.g. for the Sn market, the inventory result for the consequential 95.5Sn–3.8Ag–0.7Cu study was assumed to be identical to the one for the attributional 95.5Sn–3.8Ag–0.7Cu study. All results below for AT and WT are expressed per 530 mm^3 solder paste applied to a printed board assembly before the reflow soldering process.

The origin of AT is the usually accepted concentrations tolerated for gases, and it is expressed as the *potential* amount of air volume that will be polluted by air toxins released during the life cycle

The origin of WT is as for AT limit values. Specifically the WHO drinking water limit guideline amount for Hg^{2+} and Hg^{4+} ions is 1×10^{-9} kg/dm^3 hence the indicator 1×10^{9} dm^3/kg. The indicator is expressed as the water volume that will be polluted by water toxins.

5.3.2 Results and Discussion

The total AT results for the CLCA comparison between 62Sn–36Pb–2Ag and 95.5Sn–3.8Ag–0.7Cu were 3,030 m^3 and 2,320 m^3, respectively. The total AT results for the ALCA were 3,060 m^3 and 2,320 m^3, respectively. Below in Table 5.4 it is shown that Pb emissions to air from incineration of printed board assemblies (PBA), together with SO_2 emissions from electricity and Sn production, were the hot-spots in this study.

Table 5.4 Top contributors to AT given per solder paste type

Solder paste	Life cycle stage	Unit process	Flow	% Contr.
62Sn–36Pb–2Ag	End-of-life	Solder incineration	Pb emissions to air	34
	Manufacturing	Sn production	SO_2 emission to air	23
	Solder paste application	Electricity production	SO_2 emission to air	18
95.5Sn–3.8Ag–0.7Cu	Manufacturing	Sn production	SO_2 emission to air	44
	Solder paste application	Electricity production	SO_2 emission to air	26

In Fig. 5.7 are shown the consequences for AT.

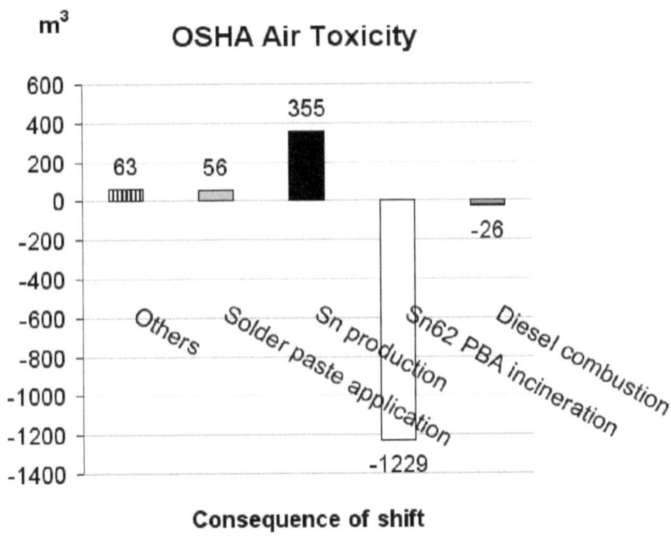

Fig. 5.7 The consequential AT results obtained when subtracting the CLCA 62Sn–36Pb–2Ag from the ALCA (=CLCA) 95.5Sn–3.8Ag–0.7Cu

The consequence is that the total AT will decrease by 780 m^3, or by 25vol%. The total WT results for the CLCA comparison of 62Sn–36Pb–2Ag and 95.5Sn–3.8Ag–0.7Cu were 2,008 dm^3 and 16 dm^3, respectively. The total WT results for the ALCA were 4,510 (75,800 extreme case) dm^3 and 16 (2,000 extreme case) dm^3 respectively. For the extreme case it was assumed that the fractions leached from 62Sn–36Pb–2Ag were 100wt%.

From incineration of solder 27.5wt% ("fumes" and "fly ash") is emitted to air and the rest landfilled with 100wt% emitted to water.

The latter represents the extreme case which only shows the leaching potential. Nobody can tell exactly what will happen in the end-of-life or in the future.

Almost 0wt% can also be emitted if a short enough time is chosen. 0.1wt% for solder metals is a suitable fraction for a foreseeable and stable period.

Observe that all values used in Tables 5.1–5.3 for landfill and incineration water emissions of Pb, Ag and Sn have been changed according to above rules.

In Table 5.5 a summary of the most important contributors is shown.

Table 5.5 Top contributors to WT given per solder paste type

Solder paste	Life cycle stage	Unit process	Flow	% Contr.
62Sn–36Pb–2Ag	Manufacturing	Pb production	Pb emissions to water	64
	Manufacturing	Pb production	Cd emissions to water	30
	End-of-life	Solder landfill	Pb emissions to water	3
95.5Sn–3.8Ag–0.7Cu	Solder paste application	Electricity production	Chloride to water	22

The total WT results (moderate leaching scenario 0.1 wt% all of solder metals) for the CLCA reflect that Pb is more or less eliminated from the solder life cycle. In Fig. 5.8 the consequences for WT are shown.

The consequence is that the total WT will decrease by 1,990 dm^3, or by 99vol%.

The CLCA results indicate that the overall AT will decrease.

The most important differences between 62Sn–36Pb–2Ag and 95.5Sn–3.8Ag–0.7Cu are several. The Pb emissions released from incineration of printed board assemblies are most significant due to the AT factor for Pb compared to Ag and Sn. SO_2 emissions from Sn and energy production are also important.

The avoidance of aforementioned Pb emissions and NO_x emissions from diesel combustion, lower the total AT *more* than emissions from increased Sn production and energy production.

Pb and other heavy metal emissions from electricity production could help underestimate the AT.

The CLCA results indicate that the overall WT will decrease considerably. The most important difference between the solders is the Pb production. The Pb and Cd emissions dominate, especially from long-term slag disposal. The leaching can be almost 0wt% or 100wt% depending on time horizon, unexpected events, and type of landfill. The present CLCA study predicts that the Pb emissions to air and water will decrease.

For ALCA this was obvious, but not for the CLCA. The Pb emissions from alternative Pb usages are hypothetically lower than from solder incineration. Pb is most notable for WT and the consequences of the shift are favorable as shown in Fig. 5.8. Moreover, WT for electricity production could be of higher importance, but not likely to change any conclusions. The data quality is poor in the model of the scrap Pb market and waste management of competing sources of scrap Pb. This part of the model does not have a big influence on the results.

The present study cannot with ease be compared to the one by USEPA [3], among other things due to inventory intransparency. The two most similar solder pastes, compared by USEPA using ALCA, to the present ones were 63Sn–37Pb (Sn–Pb) and 95.5Sn–3.9Ag–0.6Cu. The toxicity categories used by USEPA are more diversified, e.g., “occupational non-cancer” and “public human health,” than AT and WT. However, “aquatic ecotoxicity” is similar to WT and therefore the ALCA results could be compared. E.g., for ALCA USEPA the “aquatic ecotoxicity” hotspots within the life cycle for SnPb solder paste were emissions of Pb-ions to water, just as for the present WT study.

It is uncertain how much of the Sn, Ag, Pb and Cu air emissions eventually will be water toxins.

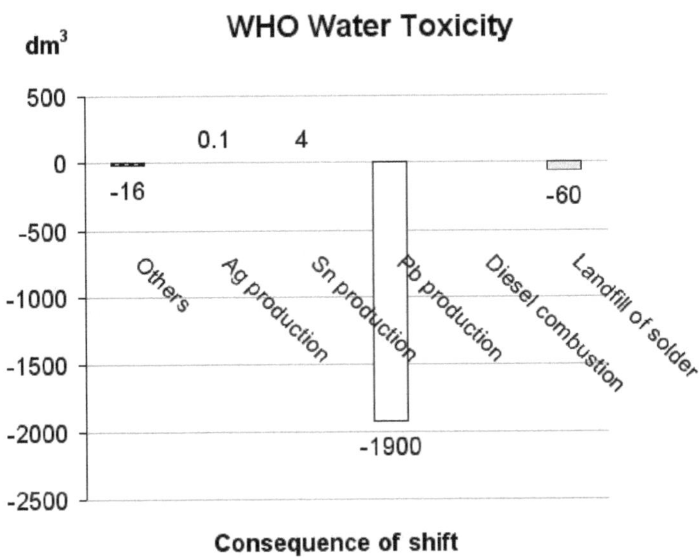

Fig. 5.8 The consequential WT results obtained when subtracting the CLCA 62Sn–36Pb–2Ag from the ALCA (=CLCA) 95.5Sn–3.8Ag–0.7Cu

That research would especially be important for Pb and Ag. Comparing the ALCA results for 62Sn–36Pb–2Ag and 95.5Sn–3.8Ag–0.7Cu is not the same as predicting the consequences of shifting from 62Sn–36Pb–2Ag to 95.5Sn–3.8Ag–0.7Cu. Occasionally, as seems to be the case regarding the solders in the present study, the two techniques, ALCA and CLCA, more or less provide the same conclusion. In any LCA study there is always room for improving the overall models and the precision of the input data.

5.3.3 Conclusions

The shift from 62Sn–36Pb–2Ag to 95.5Sn–3.8Ag–0.7Cu solder paste is likely to result in reduced air toxicity (AT) and water toxicity (WT). For AT, a significant increase was detected due to increased Sn production and generation of electricity for solder application, but this increase was offset by a decrease mainly due to avoided incineration of printed board assemblies containing Pb.

Contrary to the case for global warming [1] and AT, no significant increases in WT could be related to the increased Sn production and generation of electricity for solder application. The WT result would be highly dependent on the Pb leachability. Identified by CLCA methodology, the alternative use of Pb (storage batteries) will offset the WT reduction related to Pb production to about 50%.

5.4 Future Solders Assessed

The purpose of this case study was to explore the potential environmental effects associated with Ag-free solder pastes. The starting point was the Ag-based solder alloy 95.5Sn–3.8Ag–0.7Cu as used by the Japanese society. The potential environmental effects of the Ag substitution are explored for a number of Ag-free solder pastes and LIME was used as evaluation method.

In combination with uncertainty for inventory data the total uncertainty can be obtained. The software Chain Management by Life Cycle Assessment (CMLCA) was used for the calculations [6]. The statistical null hypothesis *t*-test was used to determine the probability of mistakenly favoring one alternative before the other when they actually are equal [7].

By *CMLCA* the coefficients of variation, CV, were obtained by Monte Carlo Simulation and by Eqs. 5.1 and 5.2 the standard deviations were derived, and by Eqs. 5.3 and 5.4 the uncertainty factors for a 95% confidence interval in the log-normal distribution. Then the orders of magnitude of the errors and test statistic, ε, were obtained by Eq. 5.5.

Values which were used for *A*, *B*, and *CV* are shown in Table 5.6.

$$\sigma_A = \sqrt{\ln(CV_A{}^2 + 1)} \tag{5.1}$$

$$\sigma_B = \sqrt{\ln(CV_B{}^2 + 1)} \tag{5.2}$$

$$k_{0,A} = e^{1.96 \times \sigma_A} \tag{5.3}$$

$$k_{0,B} = e^{1.96 \times \sigma_B} \tag{5.4}$$

$$\varepsilon = \frac{\log_{10}\left(\frac{A}{B}\right)}{\sqrt{(\log_{10}(k_{0,A}))^2 + (\log_{10}(k_{0,B}))^2}} \tag{5.5}$$

For example, in spreadsheet software Excel®, the *t-test* of the comparison between *A* and *B* can be performed as:

```
=LOG(baseline value for A / baseline value for B) / SQRT((uncertainty for A expressed in orders of magni-tude^2 + uncertainty for B expressed in orders of mag-nitude^2))
```

ε is compared to the inverse of the *t-distribution* which in Excel is calculated as:

```
=TINV(probability which corresponds to ε,150)
```

That probability should be less than 5% for a 95% confidence interval and 33% for a 67% confidence interval. One needs around 2.8 orders of magnitude in difference (including uncertainties of one order of magnitude for both *A* and *B*, $k_{0,A} = k_{0,B} = 10$) between *A* and *B* in order to have a 95% confidence. In Sect. 4.7.2 it is further discussed how SimaPro software calculate the probability of *A* being better than *B*.

5.4.1 Description of Life Cycles

Included are metal, alloy, and powder production, use phase (stencil printing, component mounting, reflow) and the disposal stage.

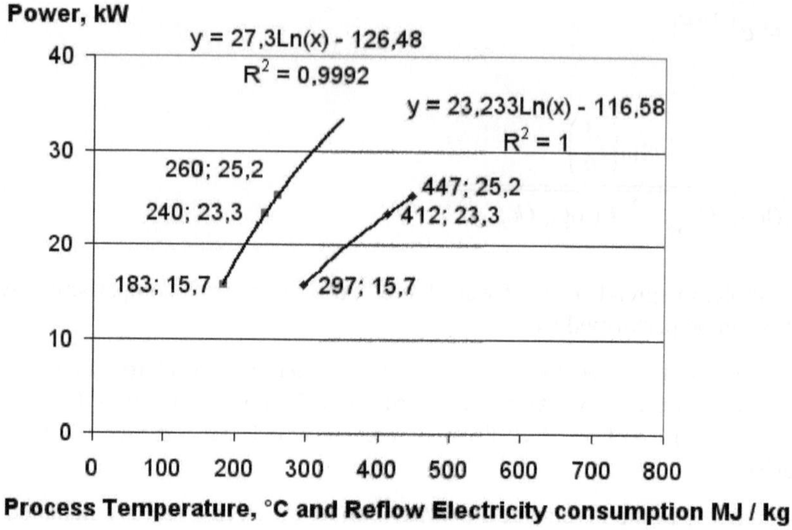

Fig. 5.9 Estimation of reflow electricity consumption for solders

A previous LCA comparing a solder paste and an ECA [5] have shown that intermediate unit processes (e.g., flux production or alloy production) are of no immediate significance.

The functional unit is *the amount of material needed to cover a PWB pad area of 10.19 cm*^{2}. This amount is around 0.56–0.6 g for the solder pastes. An identical end-of-life scenario [5] was assumed for all solders.

The present potential recycling for Ag is in total around 80wt% as metals can be recycled also when exported and when recycled by other routes than e-waste collection. Ag losses are up to 99% recycled within Sn–Ag–Cu (alloy/powder/paste) production and stencil printing.

No other metals than Ag are assumed to be recycled within the pre-metal production phase. The solder pastes consist of 90wt% metal alloys and 10wt% chemicals. At this stage no differences were assumed in pre-metal production steps (powder and paste) and material application (stencil printing and component mounting) between 95.5Sn–3.8Ag–0.7Cu and the Ag-free solder materials. The eventual differences will be explained by metal production, reflow electricity usage, density, and disposal stage toxicity. Ce and La production was approximated with Cu production. The reflow electricity usage (RE) for 95.5Sn–3.8Ag–0.7Cu was calculated as 0.447 [MJ/g] [3] × 0.0125 [cm] × 10.19 [cm^2] × 4.39 [g/cm^3] = 0.25 MJ. The RE values were obtained by logarithmic extrapolation (Fig. 5.9) from the values of 95.5Sn–3.8Ag–0.7Cu, 42Sn–58Bi and 63Sn–37Pb.

5.4.2 Results and Discussion

Table 5.6 shows the main result of the case study.

Table 5.6 Baseline CO_2 and LIME results per functional unit for solder pastes

Material	CO_2 (g)	CV %	LIME score ¥	CV %
95.5–Sn3.8–Ag–0.7Cu	70	6	1.19	120
91Sn–9Zn	63	6	1.07	79
98.8Sn–0.7Cu–0.25Ce–0.25La	67	5	1.15	132
98.9Sn–0.7Cu–0.4Co	71	6	1.19	152
99.3Sn–0.7Cu	70	6	1.18	108
42Sn–58Bi	58	6	0.81	82
89Sn–8Zn–3Bi	62	6	1.05	102

Figures 5.10 and 5.11 show the probability density functions (pdf) for the solder pastes.

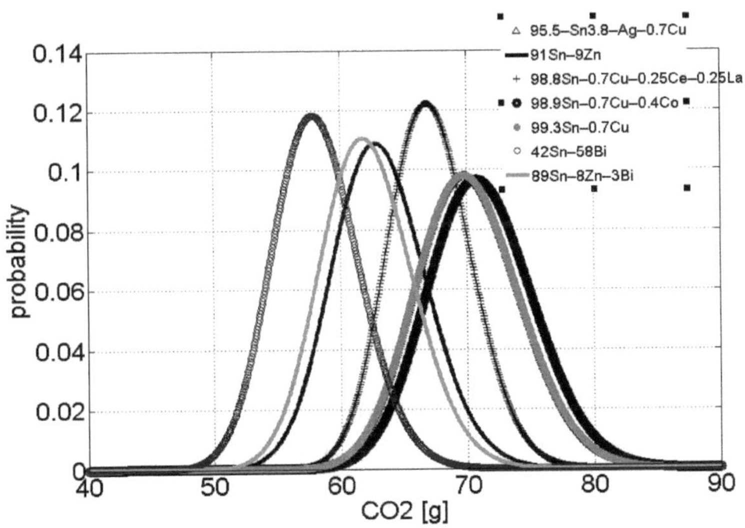

Fig. 5.10 Probability density functions for CO_2 emissions for seven solder pastes

The MATLAB® code generating Fig. 5.10 is:

```
x = (40:0.05:90);
y1 = lognpdf(x,log(70),log(1.06));%95.5-Sn3.8-Ag-0.7Cu
y2 = lognpdf(x,log(63),log(1.06));%91Sn-9Zn
y3 = lognpdf(x,log(67),log(1.05));%98.8Sn-0.7Cu-0.25Ce-
0.25La
```

```
y4 = lognpdf(x,log(71),log(1.06));%98.9Sn-0.7Cu-0.4Co
y5 = lognpdf(x,log(70),log(1.06));%99.3Sn-0.7Cu
y6 = lognpdf(x,log(58),log(1.06));%42Sn-58Bi
y7 = lognpdf(x,log(62),log(1.06));%89Sn-8Zn-3Bi
plot(x,y1,'black^',x,y2,'black-
',x,y3,'black+',x,y4,'blueo',x,y5,'blue*',x,y6,'blueo',
x,y7,'green-')
h = legend('95.5-Sn3.8-Ag-0.7Cu','91Sn-9Zn','98.8Sn-
0.7Cu-0.25Ce-0.25La','98.9Sn-0.7Cu-0.4Co','99.3Sn-
0.7Cu','42Sn-58Bi','89Sn-8Zn-3Bi',7);
set(h,'Interpreter','none')
grid
xlabel('CO2 [g]'); ylabel('probability')
```

CMLCA features seven tools for life cycle interpretation. One of them is called "Key issues" (Table 5.8) by which the contributing flows to the total uncertainty are obtained. The uncertainty of the weighting factor for Air pollution is one of the key issues as identified with *CMLCA*.

Tables 5.7 and 5.8 show the contribution analysis and key issues for 98.9Sn–0.7Cu–0.4Co.

Table 5.7 Top contributors to the LIME score for 98.9Sn–0.7Cu–0.4Co

Process	Environmental flow	Category	Value (year)	Contribution (%)
[P29] Tin production	[E54] Part. Bi. 10.[air]	[C12] Air pollution (HH)	0.125	10
[P34] Electricity, Chinese average	[E68] Sulfur dioxide[air]	[C12] Air pollution (HH)	0.124	10
[P34] Electricity, Chinese average	[E54] Part. Bi. 10.[air]	[C12] Air pollution (HH)	0.11	9
[P29] Tin production	[E53] Part. Be. 2.5 and 10[air]	[C12] Air pollution (HH)	0.108	9
[P29] Tin production	[E68] Sulfur dioxide[air]	[C12] Air pollution (HH)	0.091	8
[P33] Electricity, Japanese average	[E68] Sulfur dioxide[air]	[C12] Air pollution (HH)	0.0821	7
[P29] Tin production	[E72] Tin, in ground	[C16] Resource consumption (PP)	0.0688	6
[P33] Electricity, Japanese average	[E54] Part. Bi. 10.[air]	[C12] Air pollution (HH)	0.043	4
[P34] Electricity, Chinese average	[E53] Part. Be. 2.5 and 10[air]	[C12] Air pollution (HH)	0.0426	4
[P34] Electricity, Chinese average	[E13] Carbon dioxide[air]	[C1] Global warming (HH)	0.0422	4
All	All	All	0.837	70

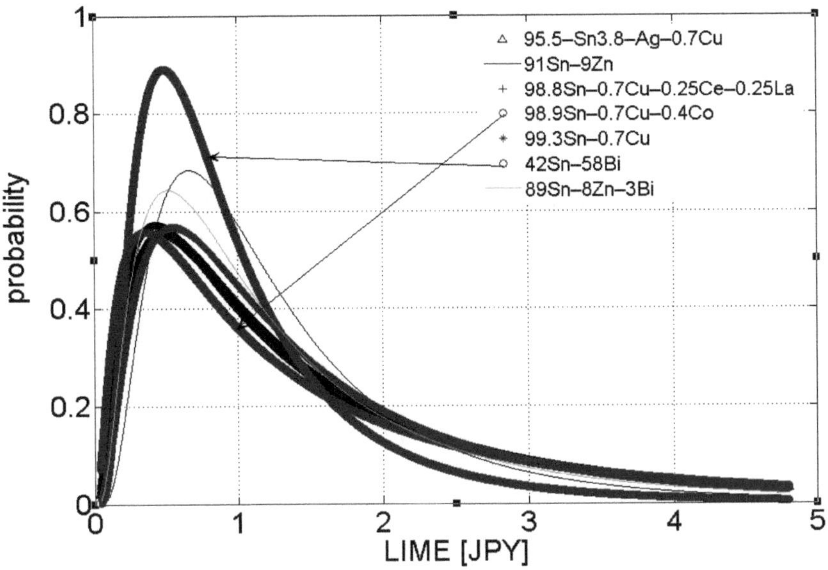

Fig. 5.11 Probability density functions for LIME scores for seven solder pastes

The MATLAB® code generating Fig. 5.11 is:

```
x = (0.04:0.0005:4.8);
y1 = lognpdf(x,log(1.19),log(2.6));%95.5-Sn3.8-Ag-0.7Cu
y2 = lognpdf(x,log(1.07),log(2));%91Sn-9Zn
y3    =     lognpdf(x,log(1.15),log(2.7));%98.8Sn-0.7Cu-
0.25Ce-0.25La
y4 = lognpdf(x,log(1.19),log(2.98));%98.9Sn-0.7Cu-0.4Co
y5 = lognpdf(x,log(1.18),log(2.4));%99.3Sn-0.7Cu
y6 = lognpdf(x,log(0.81),log(2.04));%42Sn-58Bi
y7 = lognpdf(x,log(1.05),log(2.32));%89Sn-8Zn-3Bi
plot(x,y1,'black^',x,y2,'black-
',x,y3,'black+',x,y4,'blueo',x,y5,'blue*',x,y6,'blueo',
x,y7,'green-')
h   =   legend('95.5-Sn3.8-Ag-0.7Cu','91Sn-9Zn','98.8Sn-
0.7Cu-0.25Ce-0.25La','98.9Sn-0.7Cu-0.4Co','99.3Sn-
0.7Cu','42Sn-58Bi','89Sn-8Zn-3Bi',7);
set(h,'Interpreter','none')
grid
xlabel('LIME [JPY]'); ylabel('probability')
```

Table 5.8 Top contributors to the uncertainty of the LIME score of 98.9Sn–0.7Cu–0.4Co

Flow	Process/Category	Variance (year^2)	%
[E54] Part. Bi. 10.[air]	[C12] Air pollution (HH)	0.339	25
Weight	[C12] Air pollution (HH)	0.324	23
[E68] Sulfur dioxide[air]	[C12] Air pollution (HH)	0.11	8
[G5] SnCoCu alloy	[P16] SnCoCu powder production	0.0911	7
[G15] SnCoCu powder	[P49] SnCoCu paste production	0.0923	7
[G83] SnCoCu paste	[P63] Stencil printing SnCoCu paste production	0.0947	7
[G162] Tin	[P9] SnCoCu alloy production	0.0889	6
[E53] Part. Be. 2.5 and 10[air]	[C12] Air pollution (HH)	0.0724	5
[G49] Electricity, Chinese average	[P93] Reflow oven, SnCoCu paste production	0.0468	3
All	All	1.26	91

The *t*-test (here applied to 42Sn–58Bi compared to 98.9Sn–0.7Cu–0.4Co in a LIME comparison) gives information about the size of the area occupied by overlapping curves in Figs. 5.10 and 5.11, i.e., the probability. Figures 5.10 and 5.11 would have to be adjusted to reflect the correlation between uncertainties in input data. Anyway for Fig. 5.10 the absolute overlap is

LOG(58/71)/SQRT((0.052^2+0.052^2))= ε = TINV(0.235,150)

Hence, the minimum probability for a true null hypothesis is around 23%. For all other LIME score comparisons that probability is higher. CMLCA calculates the *absolute* variation per alternative, however not *t*-tests.

Considering the large uncertainties involved, seemingly a difference of around two to three orders of magnitude is needed to be able to distinguish "A" from "B" when using an LCIA single index method as LIME as decision support. However, as the systems studied use identical, e.g., Sn and electricity production modules the difference required is smaller. By using the CMLCA software, developed by the Centre for Environmental Science at Leiden University, it is possible to simultaneously include uncertainties for inventory and impact assessment. This was done for the comparisons of the interconnection materials.

Beforehand it was expected that 42Sn–58Bi would be better than 95.5Sn–3.8Ag–0.7Cu, for both LIME and CO_2, due to its fewer input materials, lack of Ag, and lower production environmental load. This suspicion was strengthened by this screening study.

5.4.2.1 CO_2 Comparison

For solders, Fig. 5.10 shows that 42Sn–58Bi has a good chance of being better than other solders, except those which include Zn. The probability is more than 75% for 42Sn–58Bi being lower than 98.9Sn–0.7Cu–0.4Co, mostly explained by the latter's higher soldering temperature.

5.4.2.2 Resource Indications

Bi, Ce, and La lack LIME indices for the impact categories belonging to "Resource consumption" and it is likely that 42Sn–58Bi is underestimated in the present study. Ce and La currently have relatively high economic values compared to, e.g., Cu, but no safe conclusion can be drawn about the environmental impact of production. The parent ores for Ce and La are distributed so the future mining could be costly [8].

Dynamic material flow analyses are probably needed to understand the present overall metal consumption patterns. Recycling of Sn, Bi, and Ag will theoretically make sense as the interconnection materials industry will be one of the largest consumers of these metals.

5.4.2.3 Toxicity

Human toxicity and ecological toxicity for Ag, Co, and Zn are not yet covered in LIME. However, Lincoln et al. suggests that the leaching potential from *mobile phones* is as high for Co as for Pb [9]. (Sn was not evaluated by Lincoln et al. but it is generally thought that Sn has a low leaching potential as well as toxicity in the present circumstances.)

In a waste extraction test (WET), Ag, Zn, Cu, and Co leach on average 0.006 (toxicity characteristics leaching procedure), 52.4, 0.027, 1.21 mg/dm^3 respectively.

Assuming 1.5 wt% solder on the PWB, 100 g sample, and 2 dm^3 (~2 kg) test solution, the leaching potential for Co is ((0.00121 g Co/dm^3 × 2 dm^3)) / 1.5 g solder) = 1.16×10^{-2} g Co/g 98.9Sn–0.7Cu–0.4Co.

For the total threshold limit concentration test, Lincoln et al. concluded that Cu, Ni, Zn, and Ag are concerns.

Co, Zn, and Cu are aquatically toxic but the actual bioavailablity is not known for e-waste applications. In order to perform a very simple sensitivity test, the disability adjusted life year (DALY) values for the very toxic metal Cd for "Human toxicity" and "Ecotoxicity" were added to Ag, Cu, Co, and Zn. The result of the sensitivity check is that Zn's rather high leaching potential is noticeable and could be a problem.

The toxicity of Zn is, however, naturally much lower than for Cd, explaining why no alarming results were obtained. The check suggests that the risks for toxic poisoning are very low for the present applications.

Regarding the increases of LIME values, which were not remarkable, metal emissions from metal and electricity production processes appeared to be *more* important than emissions from end-of-life processes, of which unregulated emissions were most important.

The data sources used for Ag and especially the Bi production have rather high uncertainties for the Pb, Cd, and Hg water emissions. This is due to that data used differ between Pb and Ag regarding inclusion of the disposal of mine tailings.

5.4.2.4 LIME Comparison

Moreover, Fig. 5.11 shows that statistically small differences exist in between the most common Ag-based solder paste and some examples of Ag-free counterparts. It is very important to use representative LCI data and the LCA practitioner should as a rule use primary data. However, those are very difficult to obtain for metal production but more and more useful data are published for each year.

Reflow electricity usage and Sn production dominate the explanations for solder pastes and therefore the LIME score of 42Sn–58Bi is lower than for the others. SO_2 and different kinds of "Particulates" are often among the most significant flows in LIME evaluations. Moreover, Ag-based materials would score significantly higher, if Ag recycling is neglected.

This screening study indicates that there was a smaller difference between Ag-free and Ag-based materials than between Ag-based and Pb-based, especially for single index weighting methods.

This is due to the high Sn content which can be considered a *key environmental performance indicator* for Pb-free solders. Other indicators are reflow electricity usage, Ag content, and Bi content.

5.4.3 Conclusions

The solder paste 42Sn–58Bi seems to be more CO_2 advantageous than Zn-free materials and 95.5Sn–3.8Ag–0.7Cu. For LIME likely less certain differences exist. However, low-temperature solders should only be compared to each other and likewise high-temperature ones. That is, 42Sn–58Bi is not intended for the same application as e.g., 95.5Sn–3.8Ag–0.7Cu.

Obviously the leaching behavior of several metals needs further attention as well as the development of DALY-factors. Ce, La, Sn, and Bi are not priorities for toxicity assessment, but could be studied at a later stage. EPS 2000d [10] is an end-point methodology which as LIME includes uncertainty information, and can therefore for single score screening and *t*-testing can be carried out.

References

1. Ekvall T, Andrae ASG (2006) Attributional and consequential environmental assessment of the shift to lead-free solders. Int J LCA 11:344–353
2. Hendrickson CT, Lave LB, Matthews HS (2006) Environmental life cycle assessment of goods and services: an input–output approach. Resources for the Future Press.
3. US Environmental Protection Agency (2005) Solders in electronics: a life cycle assessment; EPA, Cincinnati, OH
4. Andrae ASG, Itsubo N, Inaba A (2007) Global environmental impact assessment of the Pb-free Shift. Sold Surf Mount Technol 19:18–28
5. Andrae ASG, Itsubo N, Yamaguchi H et al (2008) Life cycle assessment of high-temperature Japanese conductive adhesives. Environ Sci Technol 42:3084–3089
6. Heijungs R, Frischknecht R (2005) Representing statistical distributions for uncertain parameters in LCA – relationships between mathematical forms, their representation in Eco-Spold, and their representation in CMLCA. Int J LCA 10:248–254
7. Lenzen M (2006) Uncertainty in impact and externality assessments – implications for decision-making. Int J LCA 11: 189–199
8. Naumov AV (2008) Review of the world market of rare-earth metals. Russ J Non Ferr Met 49:14–22
9. Lincoln JD, Ogunseitan OA, Shapiro AA et al (2007) Leaching assessments of hazardous materials in cellular telephones. Environ Sci Technol 41:2572–2578
10. Bengtsson M, Steen B (2000) Weighting in LCA – approaches and applications. Environ Prog 19:101–109

Chapter 6
LCAs of Pb Solders vs. Conductive Adhesives

Abstract This chapter focuses on comparisons between conductive adhesives and solder pastes. First, plating technologies used by a quad flad pack component are compared using LIME. Then, the latest electric power and metal production LCI data from ecoinvent is used in earlier studies of high-temperature materials. As resource stress is one of the three pillars of life cycle product management, the third section explores the hidden resource flows by applying the ore-total material requirement model. The last section takes a look at future and Ag-free conductive adhesive material technologies. Moreover, traditional product level LCA (the relative LCAs for conductive adhesives) is scaled-up to global material flows to get a different distribution of the eco-profiles.

6.1 Comparative LIME Analysis of Plating Materials

The objective of this study was to quantify, using a screening life cycle impact assessment in a static accounting model, the main environmental loadings and effects associated with the product life cycles of surface metal plating for printed wiring board pads and leadframe terminals. The platings used are "pure" Sn and Au/Ni/Pd. The hypothesis for the present research was that Sn would have a lower LIME based environmental impact score than Au/Ni/Pd.

The problem addressed in order to falsify the hypothesis was: What is the comparative LIME score, expressed in JPY, for an LCA comparison between Sn and Au/Ni/Pd surface pad and terminal platings?

6.1.1 Scope

Table 6.1 provides a summary of the analyzed materials. For the Au/Ni/Pd and Sn platings the layer thicknesses were found in the literature [1, 2]. The area of one leadframe terminal was calculated based on a commercial reference where the measures could be identified, e.g., the terminal width 0.22 mm [3]. The pad area was calculated by inspecting the test board used in the study. The pad length was assumed to be 2 mm and the pad width 0.5 mm [3].

Table 6.1 Summary of the evaluated plating materials

Area/PWB pad	1 mm^2	1 mm^2
Pad plating material and thickness	Sn, 8 μm	Au, 3 nm, Pd, 0.02 μm Ni, 0.5 μm
Area/terminal	0.86 mm^2	0.86 mm^2
Terminal plating material and thickness	Sn, 8 μm	Au, 3 nm, Pd, 0.02 μm Ni, 0.5 μm

The simulation and modeling was carried out using the LCA software SimaPro 7.0.

Starting the surface plating life cycle the metals Sn, Pd, Ni, and Au are globally mined. The metals are produced as bars and both primary and secondary metals can be used, however due to lack of data on metals recycling only data for primary metals production were used. Bars of Pd, Ni, Au, and Sn are transported to plants where printed wiring boards (PWB) and microcircuit leadframe terminals are covered with platings. A PWB is a board on which no electronic components have been mounted.

The reflow soldering process where, e.g., microcircuits, having surface-plated leadframes, are placed onto the PWB and the solder is melted to form a printed board assembly (PBA). After the reflow process the PBAs are thought to be mounted inside various electronic products, which are next shipped all over the world and used.

Likely no losses of platings occur during the use phase of electronics. Used electronic devices, including solders, are in transported to landfills, incineration, or recycling, or are informally discarded.

The functional unit chosen in this study is *the amount of plating* needed for two quad flat pack(OFP)-200 components.

For both metal platings given in Table 6.1 the electricity usage, 0.03 kJ/mm^2, for plating of both pads and the terminal process was approximated by Ni electroplating electricity usage [4]. 5 mass% material losses, which were not recycled, assumed based on the process "Electroplating Nickel I" found within the ecoinvent database, which is fully integrated in SimaPro 7.0 LCA software.

All main metal production processes from cradle to gate were modeled using data for primary metal production from the ecoinvent database. After the electronic devices are considered useless they are assumed to enter one of the following four main scenarios; controlled landfills, incineration, recycling to extract the metals, or otherwise illegally disposed of. A possible distribution could be 74 mass% landfill, 14% incineration, 10% recycling, and 1% unregulated dumping. In total 43 mg Sn plating and 3.54 mg Au/Ni/Pd plating entered the waste management phase.

LIME (life cycle impact assessment method based on endpoint modeling) was used to evaluate the environmental impacts.

6.1.2 Results and Discussion

Each substance of the inventory results was (as in most LCAs) automatically compared with lists of characterization factors. Thereafter the matching substances and characterization factors were multiplied and summed for each category. Figure 6.1 shows the comparative results for ten different impacts categories.

It means that the application of 43 mg of Sn plating could lead to a social cost of around 0.5 JPY, and that the 3.5 mg Au/Ni/Pd plating indicates a cost of about 1.7 JPY. Eco-Indicator '99 is a European impact assessment method attempting to determine the interaction between products and the European environment. Eco-Indicator '99 (H) damage category Human Health can be compared to the corresponding one for LIME as both are measured in Disability Adjusted Life Years (DALY).

As shown in Fig. 6.3, one of the main similarities is that sulfur dioxide is one of the main contributors. Pb emissions factors are included in LIME "human health" but not yet in Eco-Indicator '99 "human health". The LCA perspective of environmental impacts can be likened to that of a helicopter cruising above a landscape.

The helicopter crew can study the landscape roughly from the cockpit and see which valleys and trees are high and which ones are low. However, in order to have a detailed view they have to use land surveillance equipment on the ground.

The lowest probability for a true null hypothesis as far as the characterization results should be for "acidification" and "air pollution". The end-of-life phase is inherently uncertain both regarding recycling and emissions; however, in the present study it was negligible.

In Fig. 6.2 the results for the overall LIME score are shown. Regarding characterization, as shown in Fig 6.1, Au/Ni/Pd is worse in nine of eleven impact categories mainly due to Pd production impacts, but still not in two resource impact categories explained by Sn mining.

6.1.3 Conclusions

Literature based LCA has been performed comparing two different plating materials of printed wiring board pads and component leadframes. The plating materials are dominated by Pd driving a 12 times less mass-intensive technology, Au/Ni/Pd, to be worse than Sn.

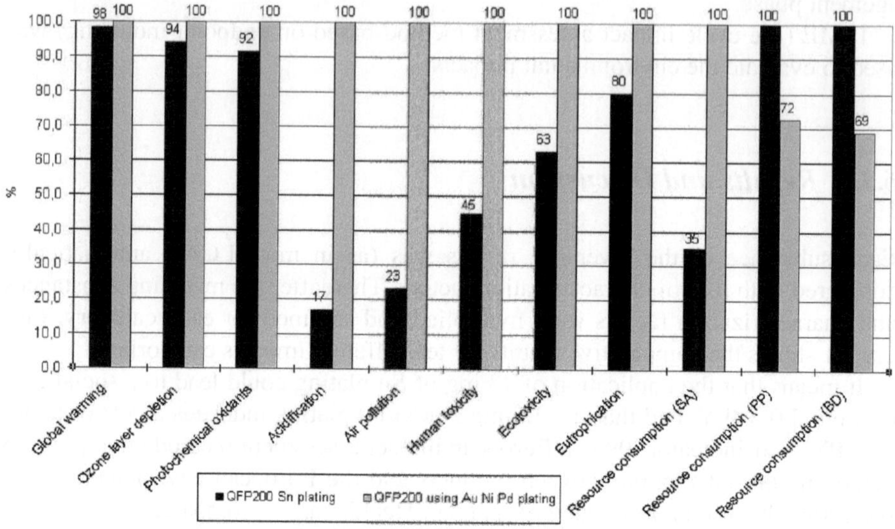

Fig. 6.1 Characterization results for platings

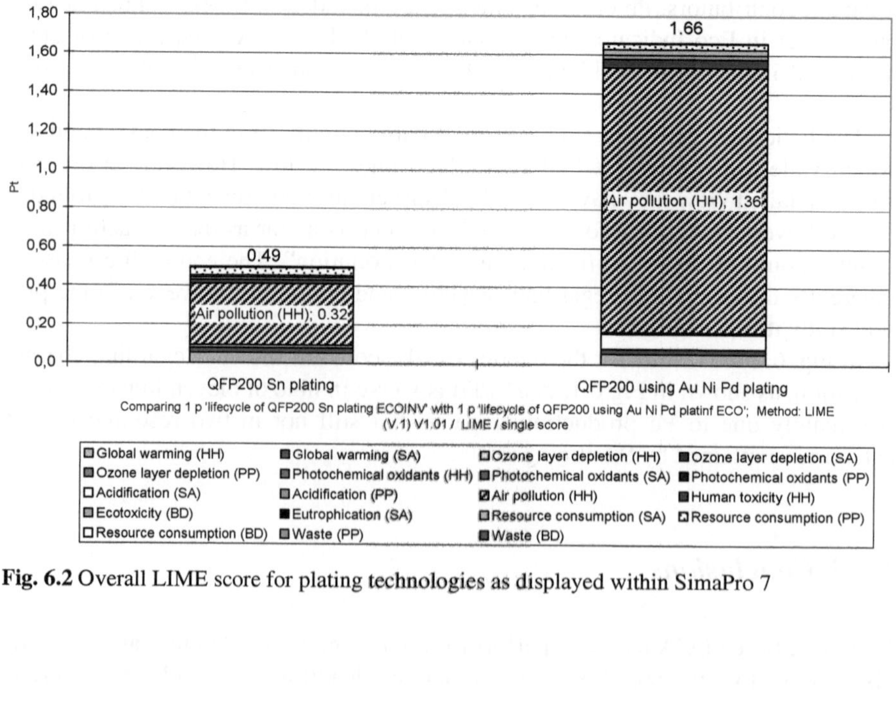

Fig. 6.2 Overall LIME score for plating technologies as displayed within SimaPro 7

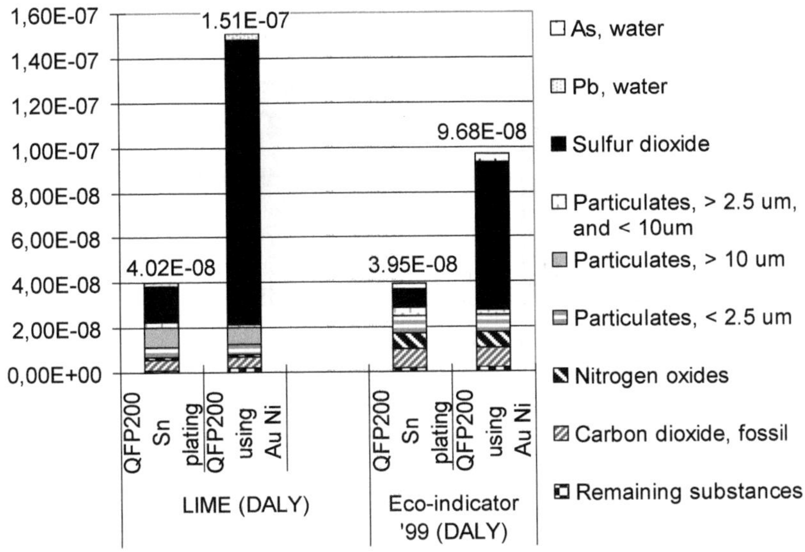

Fig. 6.3 Overall human health (DALY) scores for plating technologies for LIME and Eco-Indicator '99 (H). H = hierarchist

6.2 LCA for High-temperature Materials Using Alternate LCI Data

The study from 2008, performed in SimaPro 7.0, comparing high-temperature interconnection materials [5] showed that the results are highly dependent on the value of LIME indices for Pb emissions and the quality of the metal production LCI data.

As new LCI data for Bi and Ag production become available, the sensitivity of the results needs to be checked. A test was performed where the ecoinvent database from 2008 was scanned for more recent data which could replace some of the data used in [5]. The LCI data found were Ag production as well as Chinese and Japanese electricity mixes. These processes had previously been designed in a different manner described in [5]. Moreover, the Pb production used in [5] for 10Sn–90Pb solder paste was changed to ecoinvent data. The question asked is whether the conclusions of [5] would change if these alternate ecoinvent data were substituted for the existing data. The LIME results expressed per damage category are shown in Fig. 6.4. For Bi production no alternate ecoinvent data were found and that could partly explain the unnoticeable changes for 42Sn–58Bi–ICA. Moreover, the results must be evaluated using the so-called LIME version 2 including updated uncertainty of the impact assessment part. The software CMLCA is available for immediate use, but perhaps SimaPro will be updated with the possibility to add uncertainty information for methodologies like LIME.

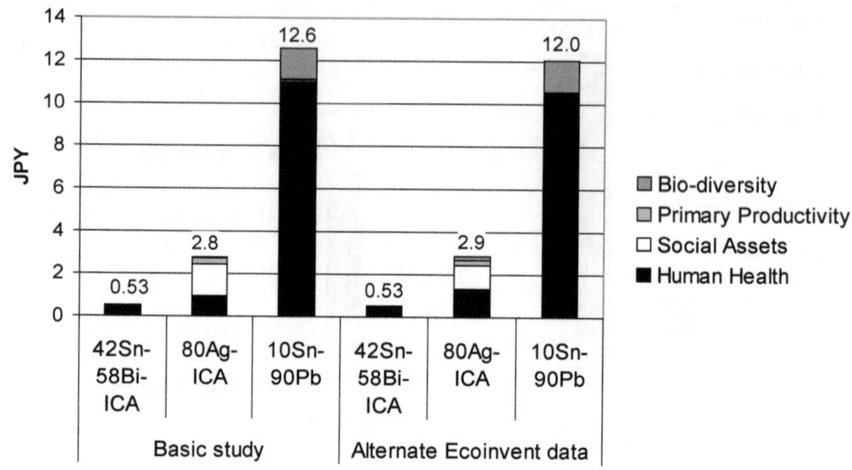

Fig. 6.4 Overall LIME scores for high-temperature materials expressed per damage category

The overall consequences of a possible shift to 80Ag–ICA and 42Sn–58Bi–ICA could be relevant to study using dynamic CLCA techniques. Simulink® within MATLAB® could be appropriate for such simulations [6]. The relevance of CLCA modeling for the above systems would be highly dependent on how much of the global Ag and Bi production that will be used to manufacture the 80Ag–ICA and 42Sn–58Bi–ICA, respectively.

The environmental research is at a stage where it is possible to "dismiss" certain steps of the interconnection life cycle as not relevant. Those are alloy and powder processes and the production of the material itself. Others are the energy usages in landfill, incineration, and illegal dumping processes. Transportations modes must be clarified further focusing on where aircraft transports occur as they contributed 3–8% to global warming. It will be possible to develop simplified tools, so-called key environmental performance indicators, based on previous life cycle simulations.

6.3 Hidden Material Flows

An index for resource burdens is one of the three important indices of life cycle management of a product. The other two are indices for global climate change and chemical pollution. Harada et al. recently developed the total material requirement (TMR) method which expresses the amount of hidden material flows by extraction of lithospheric and ecospheric resources [7].

Table 6.2 shows selected ore-TMR values calculated in 2006 by Halada and Ijima at the National Institute for Materials Science in Tsukuba in Japan.

Table 6.2 ore-TMR values for selected elements and materials

Material	TMR	Unit
Ag	4,756	kg/kg
Al	48	kg/kg
As	29	kg/kg
Au	1,113,984	kg/kg
B	138	kg/kg
Ba	512	kg/kg
Be	2,535	kg/kg
Bi	177	kg/kg
Br	1,500	kg/kg
Ca	90	kg/kg
Cd	7	kg/kg
Ce	2,015	kg/kg
Cl	52	kg/kg
Co	607	kg/kg
Coal, unspecified	12	kg/kg
Cr	26	kg/kg
Cs	11	kg/kg
Cu	360	kg/kg
Dy	9,000	kg/kg
Er	12,000	kg/kg
Eu	20,000	kg/kg
F	210	kg/kg
Fe	8	kg/kg
Ga	14,151	kg/kg
Gd	10,000	kg/kg
Ge	123,825	kg/kg
Hf	10,023	kg/kg
Hg	2,000	kg/kg
Ho	25,000	kg/kg
I	44,629	kg/kg
In	4,474	kg/kg
Ir	399,811	kg/kg
K	54	kg/kg
La	3,113	kg/kg
Li	1,483	kg/kg
Lu	45,000	kg/kg
Mg	71	kg/kg
Mn	14	kg/kg

Table 6.2 continued

Material	TMR	Unit
Mo	751	kg/kg
Nd	3,000	kg/kg
Ni	262	kg/kg
Nb	636	kg/kg
Oil, crude	7	kg/kg
Os	541,748	kg/kg
P	224	kg/kg
Pb	28	kg/kg
Pd	812,859	kg/kg
Pr	8000	kg/kg
Pt	524,823	kg/kg
Ra	2.8×10^{8}	kg/kg
Re	20,004	kg/kg
Rh	2,299,773	kg/kg
Ru	78,964	kg/kg
Rb	133	kg/kg
S	521	kg/kg
Sm	9,000	kg/kg
Sb	42	kg/kg
Se	70	kg/kg
Si	34	kg/kg
Sn	2,500	kg/kg
Na	49	kg/kg
Sr	505	kg/kg
Ta	6,815	kg/kg
Te	270,000	kg/kg
Tb	30,000	kg/kg
Tl	434	kg/kg
Th	9,011	kg/kg
Tu	40,000	kg/kg
Ti	36	kg/kg
U	22,007	kg/kg
V	1,546	kg/kg
W	191	kg/kg
Yb	12,000	kg/kg
Y	2,721	kg/kg
Zn	36	kg/kg
Zr	546	kg/kg

Ore-TMR expresses the amount of hidden materials flows by extraction of natural resources from the litho- and eco-sphere. The "hidden material flow" is the portion of the total material requirement that stays out of the economy. That is, Halada's ore-TMR model helps estimate the effective material flow rather than the conventional weight flow.

TMR considerations also highlight the rare metals content in products, giving guidelines to increase resource productivity. Perhaps the most important advantage of TMR calculations is to be able to compare any products which include various values of resources with the same resource indicator. According to ancient Chinese wisdom, resource conservation ought to be one of three important parameters of sustainability: 天地人 (tian, di, ren: heaven, earth, and people).

6.3.1 Scope

This study estimated the TMR score for the materials reported in [3]. In the life cycle inventory analysis the energy and raw material requirements, air emissions, waterborne effluents, and solid wastes are quantified per functional unit. TMR is only concerned with raw material flows. The calculations were performed by *CMLCA*.

6.3.2 Results and Discussion

In Table 6.3 a few selected inventory results are shown for all three interconnection materials.

Table 6.3 Selected inventory results for two adhesives and one solder paste given in gram per functional unit

	42Sn–58Bi–ICA	80Ag–ICA	10Sn–90Pb
Input [g]			
Coal (r)	12.4	26.7	11.1
Oil (r)	1.55	6.43	1.25
Ag (r)		0.23	
Bi (r)	0.11		
Sn (r)	0.08		0.057
Pb (r)			0.52

In Figs. 6.5 and 6.6 are shown the TMR single total scores and the substance contributions, respectively.

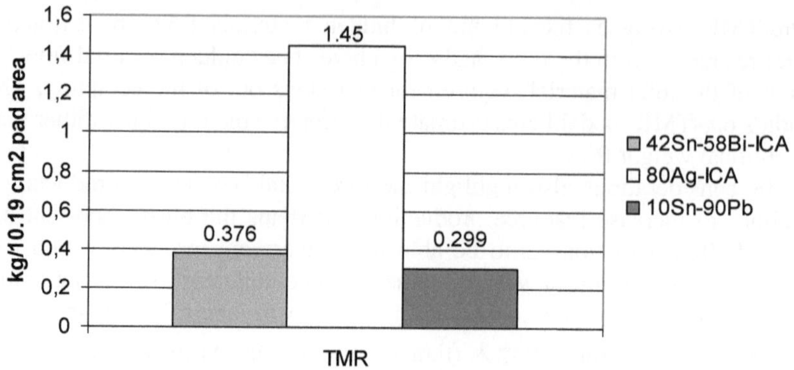

Fig. 6.5 Total ore-TMR scores for high-temperature interconnection materials

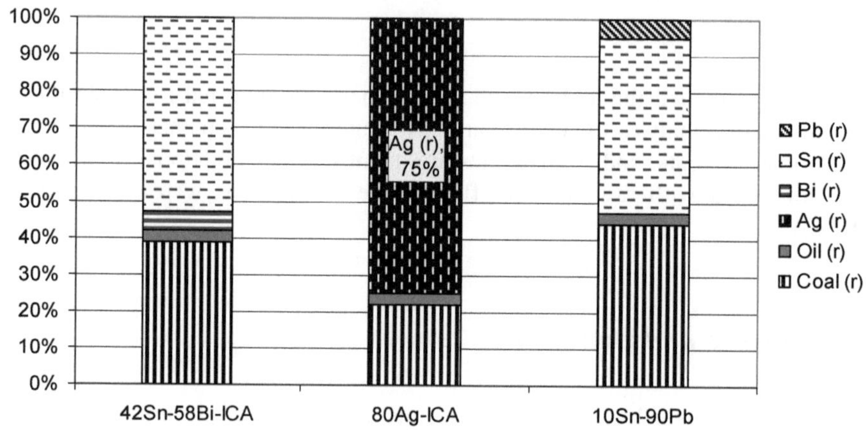

Fig. 6.6 ore-TMR contributions for high-temperature interconnection materials

It means that the application of about 0.6 g of 10Sn–90Pb could lead to a social cost of around 0.3 Yen, and that the 0.46 g 80Ag–ICA is near 1.4 Yen mainly due to the ore-TMR value of Ag.

80Ag–ICA is more resource demanding than 10Sn–90Pb and 42Sn–58Bi–ICA, however, the difference between the latter two seems to be small.

Ag is likely the metal embodying the highest overall environmental load compared to other materials which also have been suggested as fillers (e.g., silicones, metal-coated polymers and Ni) for ECAs. It is an open question if Au-coated polymer fillers are more resource intensive than Ag.

As for LIME, it does not focus on raw material requirement as much as human health costs and precious-metal cost.

Figure 6.7 illustrates the outcome of a preliminary uncertainty analysis based upon Monte Carlo simulation, input uncertainty for most inventory flows, as well as +/– 10% uncertainty range (L(0.095) in CMLCA) for or TMR-values.

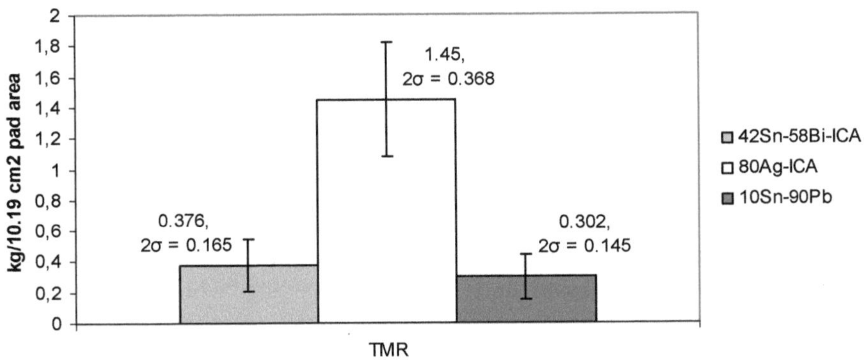

Fig. 6.7 Absolute uncertainty ranges for ore-TMR s for high-temperature interconnection materials

The possible values of Sn resources used and the ore-TMR indicator for Sn are the main reasons for the relatively high uncertainty of the 10Sn–90Pb and 42Sn–58Bi–ICA scores. Bear in mind, however, that the absolute uncertainties do not tell much about the probability of one alternative being better than the other. The reason is that, e.g., 10Sn–90Pb and 42Sn–58Bi–ICA use identical Sn data.

6.3.3 Conclusions

A screening life cycle assessment has been performed comparing three different interconnection materials. 80Ag–ICA shows signs of more hidden material flows compared to 10Sn–90Pb and 42Sn–58Bi–ICA for which the existence of a clear difference is in doubt.

6.3.3.1 Outlook

An LCA result can be validated by applying more primary data and new modeling methods, especially uncertainty analysis and dynamic modeling. Parts of the present study are based on various data from literature sources which might deviate from primary data.

More accurate data describing curing time and reflow time and the related power usage, as well as alternate models describing global metal production (especially Ag), are the most important issues in the improvement of the inventory analysis of the present materials. Uncertainty ranges need to be developed for the entire framework of ore-TMR, depletion acceleration coefficient, and resource scarcity index.

6.4 Estimation of the LIME-based Environmental Cost of Future Conductive Adhesives

Environmental impact predictions definitely belong to holistic science in contrast to analytic reductionism. This is because the environmental systems are complex and, e.g., the LCA approach demands multidisciplinary thinking which needs the knowledge and methodology from more than one discipline. This case study aimed at estimating the environmental cost of the future usage of conductive adhesives. The environmental cost will be estimated using the "willingness to pay" approach as defined in LIME.

6.4.1 Scope

An estimation of the current ECA niche usages is about 0.4 Gg based on 2.5wt% of ECA of the total global industrial Ag usage and some likely shares for different types of ECAs [5]. Japan uses around 16wt% of all BMs and it is probable that the usage will increase despite miniaturization trends, thereby implying considerable increases in the global Sn and Ag consumptions [8]. Moreover, it is possible that ECA could replace solder pastes for certain applications. For ECAs, e.g., as oxidation problems are overcome, materials such as Cu, Ni, and AlN can replace Ag [9–11]. Also Ag-based ECAs can replace Pb-based high-melting-point die-attach solder paste [5]. The LCA profiles will reflect the global distribution of conductive usage to estimate the societal environmental cost. Uncertainty analysis and hypothesis testing will be explored for CO_2 emission and weighting index comparisons.

The starting points is the 80wt% Ag particles and 20wt% epoxy-based ECA (80Ag–ICA). The potential environmental implications of the Ag substitution are explored for three Ag-free conductive adhesives. LIME was used as the evaluation method for the ECAs.

LIME is an example of a robust LCIA framework for Japan containing the relevant environmental impact categories, as well as uncertainty information. The main advantage of using LIME is that the relative importance of different environmental problems in Japan can be quantified. The main drawback could be the large uncertainty for the single score occasionally making it different to separate two comparables. This single score is a monetary value (environmental cost) which can be interpreted as the price which the Japanese society is willing to pay, e.g., to avoid the emission of cadmium into the environment. The price to sustain the environment is much more difficult to quantify than the willingness-to-pay price.

However, in combination with the uncertainty for inventory data the total uncertainty can be obtained for the single score and the software Chain Management by Life Cycle Assessment (CMLCA) and MATLAB® were used for the calculations [12]. The common statistical null hypothesis *t*-test could be used to determine the probability of mistakenly favoring one alternative over the other when they actually are equal [13]. The procedure is fully described in reference [5]. Included are metal, alloy, and powder production, use phase (stencil printing, component mounting, reflow) and the disposal stage as shown in Fig. 6.8 for 80Ag–ICA.

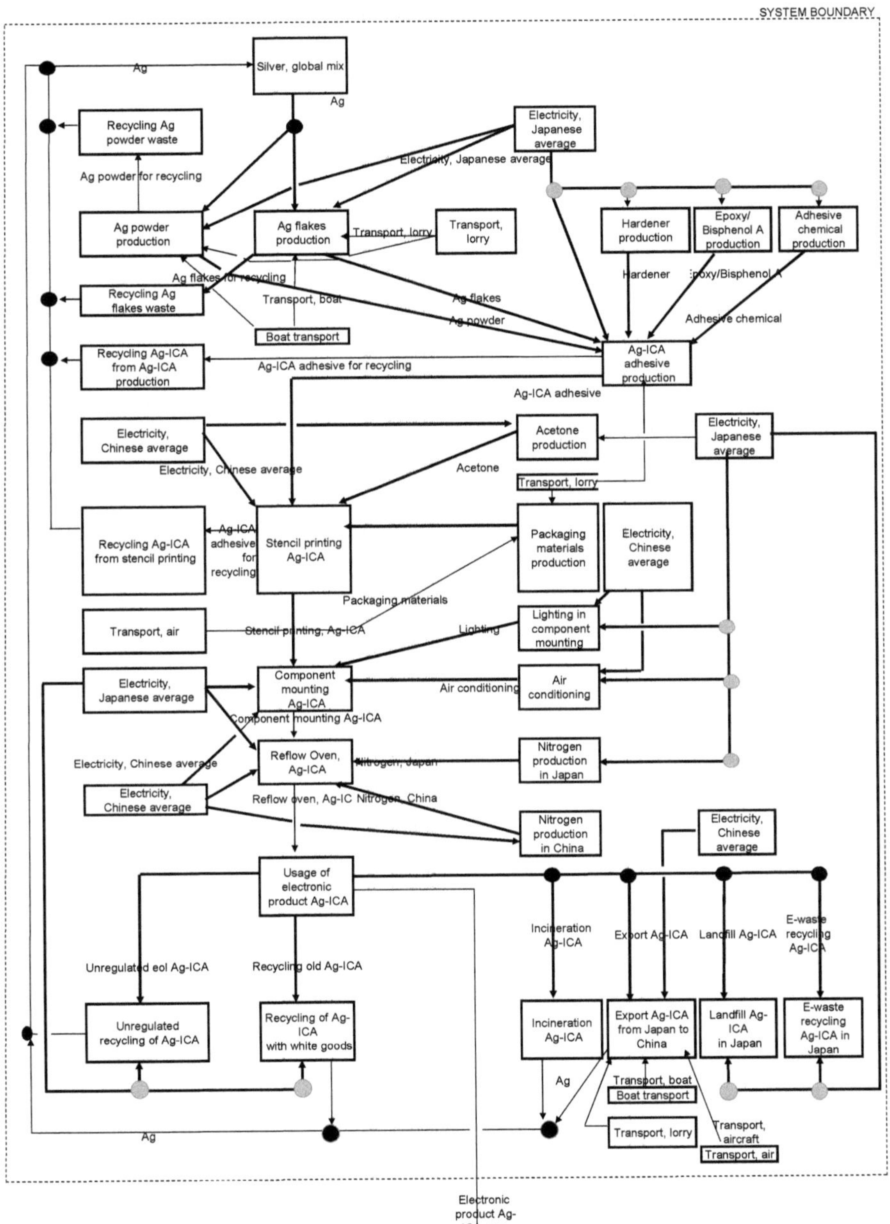

Fig. 6.8 Life cycle flowchart of 80Ag–ICA

Table 6.4 shows the foreseeable material flow amounts of conductive adhesives in the world.

Table 6.4 Expected global usage of selected electrically conductive adhesives

	2008, 80Ag–ICA dominates	Growth for ECA usage by 2010	Around 2020: 80Ag-ICA replaces 9wt% of 10Sn–90Pb solder paste, Ag–free ICAs replace 80Ag–ICA elsewhere
80Ag–ICA	~450	~630	~1,300
80Cu–ICA	~20	~35	~210
80Ni–ICA	~20	~21	~65
80AlN–ICA	~10	~14	~530
TOTAL ECA USE	~500	~700	~2,100

The rough estimations given above in Table 6.4 are based on an increase in adhesives usage from 500 to 800 tonnes between the years 2008 and 2020, that high-melting-point 10Sn–90Pb will be 12wt% of the electronic solder alloy usage, solder paste will be 9wt% of the solder alloys, and 15% lower material density and printing height for 80Ag–ICA compared to 10Sn–90Pb.

Around 1,300 tonnes of 10Sn–90Pb solder paste could as such be replaced by high-temperature 80Ag–ICA, and the rest of 800 tonnes are distributed between 80AlN–ICA 66wt%, 80Cu–ICA 26wt% and 80Ni–ICA 8wt%.

Table 6.5 shows characteristics important for LCA for the present BMs.

Table 6.5 Selected characteristics relevant for LCA of conductive adhesives

Conductive adhesive	~Density (g/cm^3)	~PT (°C)	~RE (kWh/board)
80Ag–ICA	4.3	160	0.022
80Ni–ICA	4.1	160	0.02
80Cu–ICA	4.1	160	0.02
80AlN–ICA	2.5	160	0.02

PT = Process Temperature. RE = Reflow oven Electricity consumption

The functional unit is *the amount of material needed to cover a PWB pad area of 10.19 cm*^{2} (457, 437, 437, and 267 mg respectively for 80Ag–ICA and so on). For the overall estimation is the amount of ECAs needed to perform interconnections on "all" PWBs produced by global manufacturers.

An identical end-of-life scenario was assumed for all adhesives for which the detailed assumptions are found in [5]. Cu, Ni, and Al are assumed to be recycled as Ag is within the post-metal production phase.

The Japanese and Chinese electricity mixes used are the same as in previous research [5].

Table 6.6 below shows indicators for the metals used.

Table 6.6 Baseline average CO_2, Cumulative energy demand (CED) and LIME results for selected bonding metals

Metal	~CO_2 (kg/kg)	~CED (MJ/kg)	~LIME score (JPY/kg)
Cu	2.9	64	1,300
Ag	242	4,300	10,500
Ni	24	190	1,800
Al	10	210	180

The Ag-free adhesives are assumed to consist of 80wt% metals and 20wt% epoxy. At this stage no differences were assumed in materials production and usage between the ECAs. The eventual differences will be explained by metal production, density and disposal stage toxicity.

Based on previous conductive adhesive research [1] the amount of reflow electricity per board (PBA) was 393.6 [kWh/8 hours]/18,000 [board/8 hours] = 0.022 [kWh/board]. The number of boards using ECAs produced per 8 hours was set as 18,000 for 80Ag–ICA and 20,000 for the others. All PBAs are more or less unique when it comes to reflow electricity usage.

6.4.2 Results and Discussion

Table 6.7 shows the baseline CO_2 and LIME results per functional unit.

Table 6.7 CO_2 and LIME results for conductive adhesives per functional unit

Conductive adhesives	~CO_2 (g)	~CV (%)	~LIME score (JPY)	~CV (%)
80Ag–ICA	90.4	14	2.8	57
80Ni–ICA	37.7	6	0.6	90
80Cu–ICA	32.9	5	0.7	117
80AlN–ICA	31.6	5	0.3	120

Figures 6.9–6.12 show the probability density functions (pdf) for the ECAs. The areas where the pdfs of, e.g, 80AlN–ICA and Ag–ICA overlap in Figs. 6.9–6.12, represent the most pessimistic probabilities of them being equal.

The MATLAB® code generating Fig. 6.9 is:

```
x = (23:0.05:120);
y1 = lognpdf(x,log(90.4),log(1.15));% 80Ag-ICA, sigma
is calculated from the CoV given by Monte Carlo Simula-
tion in CMLCA, exp(sqrt(ln(CoV^2+1))
y2 = lognpdf(x,log(37.7),log(1.06));%80Ni-ICA
y3 = lognpdf(x,log(32.9),log(1.05));%80Cu-ICA
y4 = lognpdf(x,log(31.6),log(1.05));%80AlN-ICA
plot(x,y1,'ro',x,y2,'ko',x,y3,'green+',x,y4,'blue*')
h = legend('80Ag-ICA','80Ni-ICA','80Cu-ICA','80AlN-
ICA', 4);
set(h,'Interpreter','none')
grid;
xlabel('CO2 [g]'); ylabel('probability')
```

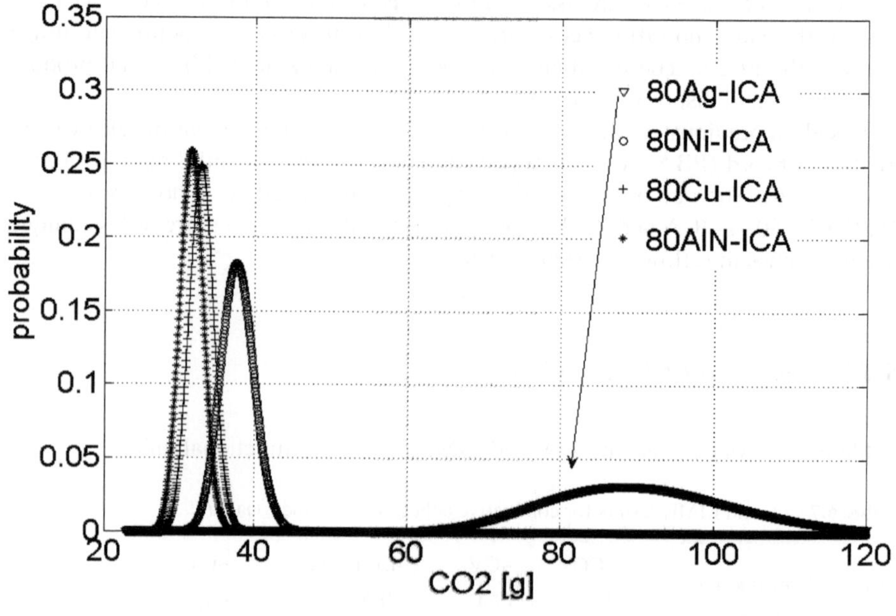

Fig. 6.9 Probability density functions of CO_2 emissions for ECAs

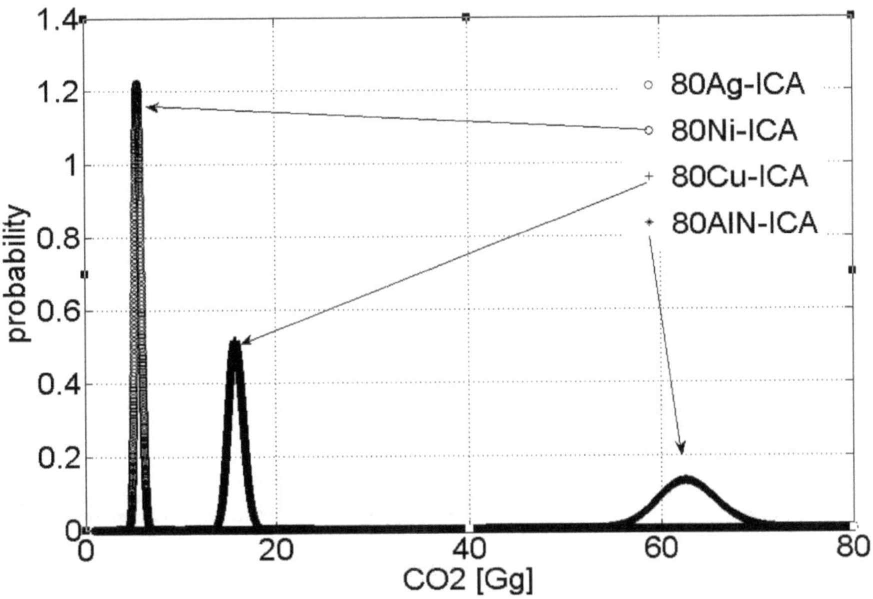

Fig. 6.10 Probability density functions of CO_2 emissions for ECAs. 3 in Table 6.4

The MATLAB® code generating Fig. 6.10 is:

```
x = (1:0.005:80);
y1 = lognpdf(x,log(257),log(1.15));%80Ag-ICA
y2 = lognpdf(x,log(5.61),log(1.06));%80Ni-ICA
y3 = lognpdf(x,log(15.8),log(1.05));%80Cu-ICA
y4 = lognpdf(x,log(62.7),log(1.05));%80AlN-ICA
plot(x,y1,'ro',x,y2,'ko',x,y3,'green+',x,y4,'blue*')
h = legend('80Ag-ICA','80Ni-ICA','80Cu-ICA','80AlN-
ICA', 4);
set(h,'Interpreter','none')
grid;
xlabel('CO2 [Gg]'); ylabel('probability')
```

80Ag–ICA is out of range in Fig. 6.10.

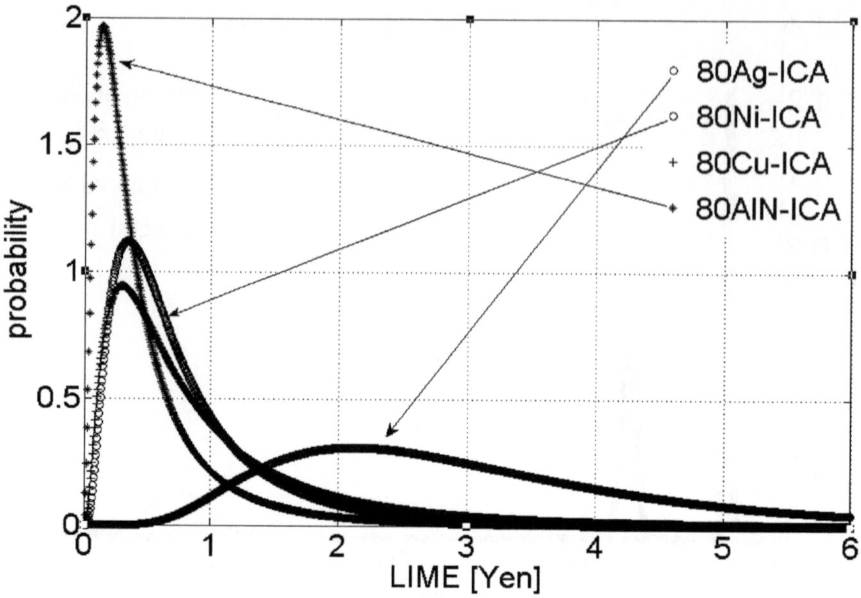

Fig. 6.11 Probability density functions of current LIME score for ECAs

The MATLAB® code generating Fig. 6.11 is:

```
x = (0:0.005:6);
y1 = lognpdf(x,log(2.81),log(1.699));%80Ag-ICA
y2 = lognpdf(x,log(0.623),log(2.16));%80Ni-ICA
y3 = lognpdf(x,log(0.702),log(2.53));%80Cu-ICA
y4 = lognpdf(x,log(0.336),log(2.57));%80AlN-ICA
plot(x,y1,'ro',x,y2,'ko',x,y3,'green+',x,y4,'blue*')
h = legend('80Ag-ICA','80Ni-ICA','80Cu-ICA','80AlN-
ICA', 4);
set(h,'Interpreter','none')
grid;
xlabel('LIME [Yen]'); ylabel('probability')
```

The MATLAB® code generating Fig. 6.12 is:

```
x = (0:0.0005:0.8);
y1 = lognpdf(x,log(7.99),log(1.699));%80Ag-ICA
y2 = lognpdf(x,log(0.105),log(2.16));%80Ni-ICA
y3 = lognpdf(x,log(0.3),log(2.53));%80Cu-ICA
y4 = lognpdf(x,log(0.668),log(2.57));%80AlN-ICA
plot(x,y1,'ro',x,y2,'ko',x,y3,'green+',x,y4,'blue*')
h = legend('80Ag-ICA','80Ni-ICA','80Cu-ICA','80AlN-
ICA', 4);
set(h,'Interpreter','none')
grid;
xlabel('LIME [GYen]'); ylabel('probability')
```

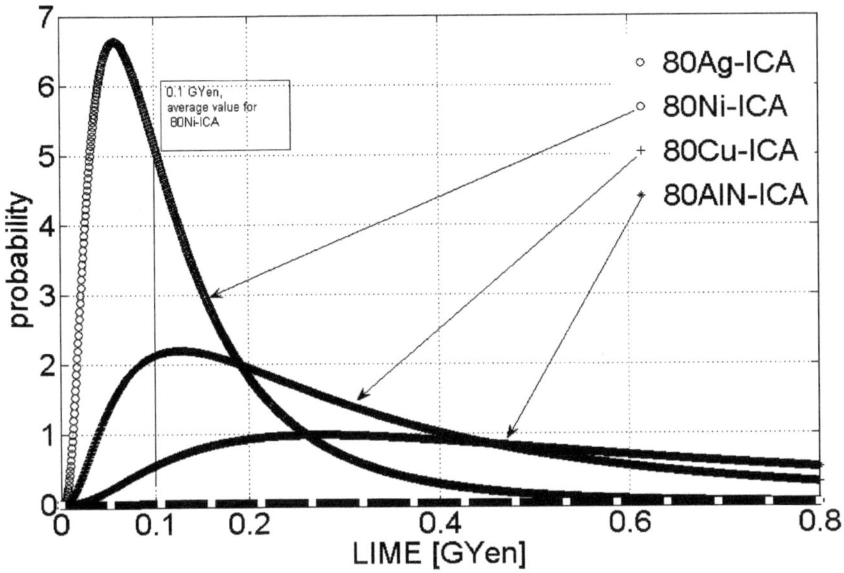

Fig. 6.12 Probability density functions of future LIME score for ECAs 3 in Table 6.4

80Ag–ICA is out of range in Fig. 6.12.

Considering the large uncertainties involved, perhaps a difference of around two to three orders of magnitude is needed to be able to distinguish "A" from "B" when using an LCIA single index method as LIME as decision support. By using the CMLCA software, developed by the Centre for Environmental Science at Leiden University, it is possible to simultaneously include uncertainties for inventory and impact assessment. This was done for the comparisons of the interconnection materials.

Beforehand it was expected that 80AlN–ICA would be better than 80Ag–ICA, for both LIME and CO_2, due to its lower density, abundance of Al, and lower production environmental load. This suspicion is strengthened by this screening study.

6.4.2.1 CO_2 Comparison

Statistically the CO_2 comparison is rather certain for the comparison between 80Ag–ICA and the alternatives. As shown in Figs. 6.9 and 6.10 the probability for a true null hypothesis should be very low for the 80Ag–ICA and 80AlN–ICA comparison explained by difference in metal production.

6.4.2.2 Resource Indications

Al was the only metal which was not contributing significantly to its resource score. Increased recycling of Ag contained in e-waste will make sense as the interconnection materials industry will rapidly increase the total global consumption of this metal.

Earlier a digital telephone was studied which printed board assembly (PBA) weighed around 160 grams and the solder around 3 grams [14]. The usage of Ag-based ECA and wave solder instead of Ag-based reflow solder could lead to that more than 0.4wt% of a PBA consists of Ag, whereas the present estimation is 0.2wt% [8].

6.4.2.3 Toxicity, Hazards, and Risks

Human toxicity and ecotoxicity factors for Ag, Ni, Cu, and Al are not yet covered within LIME. *Toxicity* is the ability of a substance to produce an unwanted effect when the chemical has reached a sufficient concentration at a certain site in the body. *Hazard* is the potential to cause harm, and the *risk* is the likelihood of harm.

The risk from a high-production-volume substance that is not very toxic and has a low hazard *may be greater* than that of a low-volume, highly toxic, and hazardous substance. It is likely that the exposures to human and biota differ in between conductive adhesives and therefore the risks are not proportional to the hazards. Below follows a discussion of these concepts in relation to the present results and the certain circumstances.

6.4.2.4 Leaching Potentials

No peer-reviewed literature was found concerning the emission behavior of ECAs. However, Muller et al. mentioned that Niederländische EinheitsNorm (NEN) test experiments demonstrated that Ag-based ECA have a low tendency of Ag leaching [15].

Several researchers have attempted to measure the relatively immediate emissions from electronics waste, in which ECAs and solders are constituents. Lincoln et al. suggests that the leaching potential from *mobile phones* is as high Ni as for Pb [16]. In a waste extraction test (WET), Ag, Cu, and Ni leached on average 0.006 (toxicity characteristics leaching procedure), 0.027, and 0.53 mg/dm^3 respectively. Assuming 1.5 wt% adhesive on a generic PWB, 100 g PWB sample, and 2 dm^3 (~2 kg) test solution, e.g., the Ni leaching potential is calculated as ((0.00053 g Ni/dm^3×2 dm^3))/1.5 g ECA) = 7×10^{-4} g Ni/g Ni-ECA. For the total threshold limit concentration test, Lincoln et al. concluded that Cu, Ni, and Ag are concerns.

Feng et al. analyzed the heavy metals in municipal solid waste incineration bottom ash by the toxicity characteristic leaching procedure (TCLP) [17]. Regardless of extraction fluid, the concentrations of Al were less than 0.01 mg/dm^3. The bottom ash used by Feng et al. is not exactly comparable to ash resulting from incineration of e-waste, but their results indicate that Al has a relatively low leachability. As for the long-term extreme case, 100wt% of the metals could reach the environmental media.

6.4.2.5 Human Toxicity, Hazard, Risk

In certain environments it is possible for hazardous materials to constantly enter the human body resulting in human toxicity. Within the LIME impact category of "Human toxicity" disability adjusted life year (DALY) values for Ni emissions values already existed. In LIME the bioavailability of the metals in Japan is taken into account. Are the potentials high or low for Ag, Cu, Ni, and Al to cause human harm when used in conductive adhesives? What is the probability that they will cause human harm?

The published DALY/kg values (Table 3.2) for metals range from 4.45×10^{-5} (Cr^{6+} emitted to air) to 4.74×10^{-2} (Pb emitted to water). In order to perform a very simple sensitivity test, the values for the "Human toxicity" of Pb were added to Ag, Cu, and Al. The result of this rudimentary sensitivity check was that the probabilities of harm are very low for the present applications.

6.4.2.6 Ecotoxicity, Hazard, Risk

Ni [18], Al [19], and Cu [20, 21] possess ecotoxic properties, but the actual bioavailablity is not yet known for e-waste applications. Are the potentials high or low for these metals to cause harm? What is the probability that they will cause harm? The LIME impact category of "Ecotoxicity" lack indices for the ECA metal emissions to air, water, and soil in the present study. The published EINES/kg values for metals range from 4.66×10^{-9} (Pb emitted to air) to 8.47×10^{-7} (Hg emitted to water). In order to perform a *very simple* sensitivity test, the values for the "Ecotoxicity"of Pb were added to Ag, Ni, Cu, and Al.

Regarding the resulting increases of LIME values, which were not remarkable, metal emissions from metal and electricity production processes appear to be *more* important than emissions from end-of-life processes of which "Export to developing countries" emissions are most important. The data sources used for upstream Ag production have rather high uncertainties for the Pb, Cd, and Hg water emissions. This is due to that data used for Ag exclude the disposal of mine tailings.

6.4.2.7 LIME Comparison – Overall Environmental Cost

Figure 6.10 shows that, statistically, there are differences in between the most common Ag-based conductive adhesive and some examples of Ag-free counterparts, both for board (PBA) level LCA and total use perspectives. Even though LCA is an initial screening tool, it is very important to use representative LCI data and the practitioner should as a rule use primary data.

However, those are very difficult to obtain for metal production, but gladly more and more useful data are published for each year. Reflow electricity usage for Ag-free ECAs and Ag production dominate the explanations for LIME in the present study. SO_2 and different kinds of "particulates" are often among the most significant flows in LIME evaluations. Moreover, 80Ag–ICA materials would score significantly higher, if Ag recycling is neglected. This screening study indicates that there is a smaller difference between PBA level LCA scores per material and functional unit (Fig. 6.11), than the differences based on overall usages (Fig. 6.12).

6.4.3 Conclusions

Statistically for CO_2 it will be advantageous to shift from 80Ag–ICA to any of the other three. For LIME the probability is high that no statistical differences exist between Ag-free ECAs. When scaling up the PBA level LCA profiles to global level material usage, the differences between CO_2 emissions for ECAs increase. For upscaled LIME scores the changes are less evident.

References

1. Kobe B, McIntyre NS (2002) Investigation of reactions between lead/tin solder and palladium surface finishes. Acta Mater 50: 4667–4676
2. Kim KS, Yu CH, Yang JM (2006) Tin whisker formation of lead-free plated leadframes. Microelectron Reliab 46: 1080–1086
3. Andrae ASG, Itsubo N, Yamaguchi H et al (2007) Screening life cycle assessment of silver-based conductive adhesive vs. lead-based solder and plating materials. Mater Trans 48:2212–2218
4. Kliopova I, Bagdonas A (2003) Optimization of electroplating processes in Lithuanian machine and instrument industry. Environ Res Eng Manag 3:29–37
5. Andrae ASG, Itsubo N, Yamaguchi H et al (2008) Life cycle assessment of high-temperature Japanese conductive adhesives. Environ Sci Technol 42:3084–3089
6. Klee H (2007) Simulation of dynamic systems with MATLAB® and Simulink®. CRC Press, Boca Raton, FL
7. Halada K, Ijima K, Katagiri N et al (2001) An approximate estimation of total materials requirement of metals. Nippon Kinzoku Gakkai-Shi 65:564–570
8. Eckelman MJ, Graedel TE (2007) Silver emissions and their environmental impacts: a multilevel assessment. Environ Sci Technol 41:6283–6289
9. Zhao H, Liang T, Liu B (2007) Synthesis and properties of copper conductive adhesives modified by SiO2 nanoparticles. Int J Adhes Adhes 27:429–433
10. Goh GF, Yu H, Yong SS et al (2005) Synthesis and cure kinetics of isotropic conductive adhesives comprising sub-micrometer sized nickel particles. Mater Sci Eng B 117:153–158

11. Yoshikawa J, Katsuda Y, Yamada N et al (2006). Electrically conductive aluminum nitride ceramics containing in-situ synthesized boron carbonitride. Key Eng Mater 317–318:653–656
12. Heijungs R, Frischknecht R (2005). Representing statistical distributions for uncertain parameters in LCA: Relationships between mathematical forms, their representation in EcoSpold, and their representation in CMLCA. Int J LCA 10:248–254
13. Lenzen M (2006) Uncertainty in impact and externality assessments: implications for decision-making. Int J LCA 11:89–199
14. Andrae ASG, Andersson DR, Liu J (2005) Significance of intermediate production processes in life cycle assessment of electronic products assessed using a generic compact model. J Clean Prod 13–14:1269–1279
15. Müller J, Griese H, Reichl H (1999) Reduced environmental impacts by lead free electronic assemblies? Proc Int Summit Lead-free Electron Assemblies IPCWorks '99, Oct. 26–28, MN
16. Lincoln JD, Ogunseitan OA, Shapiro AA et al (2007) Leaching assessments of hazardous materials in cellular telephones. Env Sci Tech 41:2572–2578
17. Feng SL, Wang XM, Wei GJ et al (2007) Leachates of municipal solid waste incineration bottom ash from Macao: heavy metal concentrations and genotoxicity. Chemosphere 67:1133–1137
18. Lock K, Janssen CR (2002) Ecotoxicity of nickel to Eisenia fetida, Enchytraeus albidus and Folsomia candida. Chemosphere 46:197–200
19. Gensemer RW, Playle RC (1999) The bioavailability and toxicity of aluminum in aquatic environments. Crit Rev Environ Sci Technol 29:315–450
20. Nor YM (1987) Ecotoxicity of copper to aquatic biota: a review. Environ Res 43:274–282
21. De Schamphelaere KAC, Janssen CR (2004) Bioavailability models for predicting copper toxicity to freshwater green microalgae as a function of water chemistry. Environ Sci Technol 38:6201–6209

Chapter 7
Discussion

Abstract This chapter contains discussion of the topics considered in previous chapters. The central questions in the debate about the ecological compatibility of Pb-free solders are given attention. Thirteen common arguments against Pb-free interconnection are penetrated. Special attention is given to the link between LCA and risk assessment from the perspective of leaching tests performed on solders and adhesives. Furthermore the concept of eco-materials is brought up and reflections are made for interconnection materials. Lastly follow there of embryos of broader theories, identified in connection to the present research, which likely will emerge.

Regarding data intensity, environmental system sciences could be similar to disciplines within sciences such as macro–micro economics and econometrics. These research fields rely on up-to-date databases in order to monitor and hypothesize the surrounding world. Environmental life cycle assessment (LCA) fills a substantial gap as an initial screening methodology and tool in environmental decision-making.

Generally, decision-making often needs decision-making support tools. The present book attempts to shed light on what LCA can contribute to comparisons between Pb-based interconnection technologies and Pb-free counterparts. The issue of a Pb-free electronics industry has many dimensions: technical, environmental, economical, and safety. LCA cannot of course answer all environmental questions about the shift to Pb-free, but this book brings up some important lessons that can be learnt from using LCA when analyzing material shifts. Though it seems Pb will gradually be banned around the world, and Pb-free soldering is technologically possible, it is interesting to understand on which environmental LCA basis the decision has been taken.

7.1 Solders

Of all interconnection materials, solders are most important to investigate. Here solders were investigated by addressing questions related to CO_2 emissions, resources and toxicity and weighted LIME scores in semiquantitative consequential LCA and attributional LCA.

The Ag LCI data used in Sect. 5.1 (10 kg CO_2/kg from cradle-to-gate) does not influence the conclusions of the CLCA. The usual Ag LCI data average is probably more than 100 kg CO_2/kg. The total ALCA (=CLCA) CO_2 score of 95.5Sn–3.8Ag–0.7Cu would increase by 5% if 100 kg/kg is used instead of 10 kg/kg.

As far as the Pb emissions, the CLCA of 62Sn–36Pb–2Ag showed that the expansion of the solder life cycle could reduce the emissions to air by 0.3wt% and the water emissions by 8wt%. The water emissions could be reduced as a result of Pb recycling which mitigates the need for primary Pb production which in turn releases Pb emissions to water. This means that the total consequence, the subtraction of two CLCAs, is very similar to the subtraction of ALCAs. The next step would be to expand the system boundaries to include coal ash disposal resulting from electricity usage (Sect. 4.1.4).

The consequential LIME calculation emphasizes the role of the social cost of Ag, despite its small material share of the solders. An important discussion is whether a consequential model for Ag and Sn markets would lead to different conclusions. How much will global Sn and Ag usage rise as far as electronics solder is concerned? Between the years 2001 and 2007 the Sn usage for e-solder has risen by 70%, but increased Sn mining has covered for it meaning that e-solders' share of global Sn production has not been dramatic. For Ag the rise has been 1,500% and here the share of Ag in e-solders has gone from 0.5 to 8% of total global Ag usage.

The shift to Pb-free solder could lead to a decreased use of Ag, Sn, and Bi in certain specific products. It is an open question which of these marginal Sn (Ag, Bi) consumers, having the possibility to substitute Sn (Ag, Bi), is most sensitive to a change in Sn (Ag, Bi) price. A change in the Sn (Ag, Bi) prices will theoretically affect the uses of these metals. For some products the Sn or Ag cost is a small part of the total production cost, while for others Sn or Ag have important functional advantages, making the demand for Sn and Ag less sensitive to changes in the Sn or Ag price. If the marginal products using Sn cannot replace Sn in the long run, the total Sn production will go up, and that is what has happened so far. It seems like all other major users of Sn, for example producers of Sn-coated cans also use the same amount of Sn after the shift to Pb-free solder. However, if the Sn price increases too much, material technologies involving Al, glass, paper, plastic, and Sn-free steel can substitute for Sn in, for example, pet food cans. On the other hand, Sn can also find new markets, for example, as alloys in automotive balance weights. The same reasoning as for Sn and Ag could be applied to other interconnection metals, most important Bi. To date not much is possible to prove regarding the links between the change to Pb-free solder and the Sn, Ag, and Bi prices shown in Figs. 4.12, 4.14, and 4.16. For a proper CLCA study, the Sn, Ag, and Bi markets should be modeled by econometricians and LCA researchers together.

The heart of the discussion for water toxicity (WT) is how much metal (especially Pb) that should be assumed to actually reach the environment from solder. For metals in solders, 0.1wt% could be a reasonable average metal loss from landfills during a foreseeable time period. The reason is that landfill leachates have historically shown small metal releases. A loss of only 0.1wt% makes the upstream emissions more important than end-of-life emissions.

The study of Ag-free solders pastes provided some new knowledge, e.g., that Ce and La are not priorities of toxicity assessments. Moreover, Sn/Ag content and reflow electricity usage are key environmental performance indictors of Pb-free solders. It is well-known that LCA is the basis for these indicators. Seen from an LCA perspective 42Sn–58Bi solder is a very good interconnection material, but it can only be used in niche applications due to its low melting point. Moreover in LCA it should only be compared to low-melting-point solders in specific applications.

Co is suspiciously leachable from mobile phones and also has some toxic characteristics. However, the usage of Co in solders is limited.

Analyses of the environmental impacts of Pb-free solders as compared to Sn–Pb solders would help avoiding worse alternatives. It is an open question why Pb-free solders have been widely introduced as to the author's knowledge no cost–benefit analysis has been published as a complement to the LCAs. The proper eco-efficiency analyses of the materials considered in this book have to wait until life cycle costs have been estimated.

It is rather remarkable that some authors argue that Ag is the largest hazard of Pb-free solders [1, 2] while others found little evidence of Ag leaching [3, 4].

7.1.1 Arguments About Pb-free Materials and Responses

A huge critique of the Pb-free materials put forward the following 13 arguments (A) against the shift which are reflected upon below.

(A1) No published evidence exists which shows the harm from Pb in electronic solder as compared to Pb in paint and gasoline.

This is not entirely true as only around 10wt% of global WEEE is recycled, despite that the technologies exist to do it right. In all fairness, how much of the *remainder* WEEE is treated in a harmful manner in, e.g., Asia and Africa? Presumably landfill dominates, and this option seems actually not too bad from a Pb-leaching perspective, but could impose a large accumulation of Pb. On the other hand, research has shown how to "eliminate" the Pb-leaching risk by cement solidification of waste PBAs [5]. The informal recycling leads to damage of health and ecosystems in the villages where it is carried out, but the extent of this occupational health problem is unknown.

(A2) The Pb-free solders containing Ag seem to increase the environmental impacts.

It depends on how end-of-life Pb emissions are accounted for in the LCA and how/if risk analysis of Pb-based solders has been made.

(A3) No publications exist proving that humans or animals have been harmed by Pb in electronic solder from normal use.

True.

(A4) Pb from electronic solders that does end up in landfills does not leach into ground water, even with years of legacy Pb from many other sources such as batteries and Pb-painted structures.

These landfill legacy facts could be a counter-argument against those researchers arguing for a rapid accumulation of Pb-containing e-waste. However, according to TCLP tests, Pb e-wastes do leach more Pb than Pb-free solders leach Ag. For Pb-based paint when tested for Pb leachability, Wadanambi et al. showed many times higher Pb concentration when leaching the paint using the TCLP than when using actual municipal solid waste landfill leachate [6]. The TCLP simulates worst-case of leaching in a landfill environment where the waste is co-disposed of with municipal solid waste. Concrete present in the laboratory leaching test dramatically reduced the Pb concentration, also for TCLP. Before, Jang and Townsend demonstrated that printed board assemblies release more Pb by TCLP than by actual Florida landfill leachates [7].

(A5) Pb-free solders contain substantial percentages of Ag (~ 3 to 4wt%), which could pose an ecotoxicological risk.

Ag does not leach much, in fact TCLP tests show a much lower leachability of Ag from Pb-free solders and conductive adhesives than Pb from Pb solders.

(A6) No economic identifiable benefits can be gained from the conversion.

The costs of the shift to Pb-free seem unnecessary but are hypothetically bearable as the solder companies are still in business and the conversion has gone rather far.

(A7) More energy, and thereby global environmental cost, is required by Pb-free soldering processes. Further, soldering machines will be degraded by increased corrosion and will have to be replaced at a faster rate.

More efficient solder ovens could offset this negative effect.

(A8) The reliability of Pb-free solders is still "in the dark" as there are numerous effects (voiding, Sn whiskers, etc.) whose causes are neither well understood nor predictable.

True.

(A9) Several LCAs comparing Pb soldering technology to Pb-free technology show that the environmental impact and cost will increase.

It is true that most life cycle energy analyses show negative effects for Pb-free technology, but that is mainly for Ag-containing materials. The functional units chosen are less of a problem than the geographical and temporal system boundaries. For Japan results are positive for Pb-free when including export and unregulated recycling, which are more negative for Pb solders. The LCA tool alone might not be able to quantify environmental risks and costs on a global perspective.

(A10) The exemptions for high-temperature solders used by, e.g., the military will create a problematic double logistic path.

The logistics argument is not convincing. No LCA has so far studied the environmental effects of changed transport patterns. Transport environmental impacts might, however, generally be under-estimated in LCA results.

(A11) Sn-Pb solders have quite few technological problems and the "old" system behavior was more or less known by researchers and engineers. The Pb-free solders were not introduced as a result of the weakness of the Sn-Pb solders. Problem solving connected to Pb-free solders will steal time and resources from other necessary innovation.

Not a strong argument as researchers would have explored both nanoscale and solder-free concepts, regardless of Pb-free legislation.

(A12) A total increasing mining of Sn resources will lead to increased rain forest depletion.

The local mining ecological effects should have been, and should be, a very important future work for sustainability assessment of material shifts in any industry.

Sn has its own mining life cycle for the most part making the LCI allocations less random than for Pb, Ag, and Bi.

(A13) The European WEEE legislation demands the recycling of electronic products, theoretically solving many of the human health risks of Pb in "old" products, and possible Sn and Ag resource depletion problems of "new" products.

Interesting argument, but why then is much e-waste still shipped to developing countries, where it is sometimes treated in a dubious manner?

It is clear that the LCAs, material flow analyses, substance flow analyses, sustainability assessments, and risk analyses need to be done *before* a material ban/shift in any industry is forced by rule of law.

Based on the assumptions in performed LCAs, it is beyond doubt that the end-of-life phase for Pb-based solders causes a higher human health hazard (potential to cause harm) than Pb-free. Therefore all conservative environmental assessment approaches suggest Pb-free.

The ecotoxicity assessment of metal release is a paramount research task. The typical LCA practice, of assuming complete emissions, is probably too simplistic. A fruitful first step suggested by Pettersson and Hertwich is to divide the metal leaching potential into three types: total metal, geoavailable metal, and highly mobile metal [8].

The hazards of Pb-free materials can be understood in the light of the waste generated by electricity production end-of-life long-term hazards/risks (coal ash leaching of heavy metals). Section 4.1.4 shows the strength of life cycle thinking for an alarm clock in discovering such eventual problems.

The often-occurring phenomenon of co-mining makes it difficult to exactly distinguish between metals environmental loadings in the raw materials acquisition phase.

These challenges of co-mining and long-term release of metals are at the center of the scientific LCA discussion on the Pb-free shift in the electronics industry.

According to the assumptions made regarding Pb leaching in landfills, they have a smaller significance than Pb emissions from unregulated recycling followed by dumping of Pb solders.

Regarding unregulated recycling in developing nations, with a high likelihood significant Pb emissions per functional unit in LCAs are released to water and soil which strongly emphasize the disposal problems.

The actual amounts emitted, and to which compartments, are dependent on the distribution of recycling procedures, and which measures are taken to prevent emissions.

An LCA of, e.g., a solder should be followed by more LCAs of the same material. These LCAs shall use the latest modeling methods and measurements.

7.2 Platings

There exist many types of component and pad platings but Sn and Au/Ni/Pd represent two quite different technologies.

Whenever Au, Pd, Pt or similar precious metals are included in a plating system, the LCA practitioner has to be very careful to get as precise as possible consumption.

The key environmental performance indicators of interconnection material plating technologies will always include Pd and similar.

Sulfur dioxide was found to be the main contributor to human health problems for both LIME and Eco-Indicator '99.

This was regardless of these methods' different scopes for the DALY/mass calculation. *LIME* and Eco-Indicator human health index coverage differ.

In the future the ReCiPe project [9] probably will enhance the understanding of how to develop new DALY indices.

7.3 Conductive Adhesives

One of the crucial questions an LCA practitioner has to ask is whether alternate data would change the conclusions of the projects. For the ALCA comparison [10] of possible environmental effects associated with 42Sn–58Bi–ICA, 80Ag–ICA and 10Sn–90Pb the relevant investigation was to test if recent published LCI data for Ag production and Japanese and Chinese average electricity would change previous conclusions. The answer for LIME comparisons was no. Bi production is still not available from the ecoinvent database and moreover, transportations need a closer look from case to case.

How to quantify resource impact is up for much debate within the LCA research community, whereby several approaches have been developed.

The so-called ore-TMR model is able to weigh different resources of an LCI result against each other. The 42Sn–58Bi–ICA, 80Ag–ICA and 10Sn–90Pb comparison [10] showed that despite rather optimistic recycling assumptions for Ag, it dominates the TMR results.

Finally for ECAs it was examined whether any conclusion would change if the functional units were magnified from PBA level to global material flow level. The differences between global material flow level LIME and CO_2 scores were larger than between usual LCA scores.

As the calculations were done in CMLCA no correct *t*-test could be done. SimaPro will be able to manage this when uncertainty for impact assessment and weighting is developed.

7.4 Are Pb-free Interconnection Materials "Ecomaterials"?

Are all Pb-free solders eco-materials? The term ecomaterial was introduced by Kohmei Halada in Japan who also developed the previously mentioned ore-TMR model [11, 12]. Ecomaterials have superior features for environmental life cycle design. In order to qualify for an ecomaterial in all senses, compared to a worse alternative, the criteria below have to be fulfilled.

- Less hazardous substances

Substance toxicity is the first consideration made by product manufacturers as the consumer will most commonly discard the products at their end-of-life. This criterion focuses on the life cycle after manufacturing. For substance toxicity Pb-free materials are preferable. This has been demonstrated by earlier work and by the case studies in the present book.

Disregarding what actually happens to the product and employing a precautions approach, the toxicity criterion alone favors Pb-free materials. However, the end-of-life hypotheses favoring Pb-free have not been falsified by actual measurements.

- "Greener" environmental profile

Materials of "green" environmental profile are those with small environmental loadings from resources mining to material manufacturing. Here the Ag-containing Pb-free materials will score worse than conventional Pb solder. The Pb emissions, from mine tailings in mines which produce both Ag and Pb, must be allocated to both.

- Higher recyclability

Recycling of materials theoretically reduces the environmental loadings associated with acquisition of virgin resources from the ecosphere, and also reduces the material flows to waste treatment. Occasionally requirements for higher material performance have made material recycling more difficult, e.g., composite recycling.

At Cu/Pb smelters it is convenient to recycle printed board assemblies using conventional Pb solders. Technology development will make it possible to recycle all kinds of Pb-free materials too.

- Higher resource productivity

The resource productivity is the material input per unit service.

LCA gives a quantitative index with which to compare the environmental impact of products or systems when the application is the same. LCA can therefore be used to calculate the resource productivity.

Materials of higher resource productivity are materials which realize higher performance of the product system (e.g., lower CO_2 emissions) with smaller resource consumption during production. Clear examples are high-tensile steels reducing the CO_2 emissions from automobiles, aluminum reducing energy usage of automotives, and magnesium reducing weight of power trains.

Performed LCAs show that the overall upstream energy consumption (and CO_2 emissions) related to the interconnection will increase as a result of the shift to Ag-containing Pb-free materials. That is, the resource productivity for CO_2 will go down.

7.4.1 Leaching of Pb – Laboratory and Landfill

In 2008 Townsend et al. measured the leachability of printed board assemblies on which the components had been connected by the solders 63Sn–37Pb, 99.3Sn–0.7Cu, 95.5Sn–4Ag–0.5Cu, 96Sn–2.5Ag–1Bi–0.5Cu, and 42Sn–1Ag–57Bi [13]. The much-referred-to toxicity characteristic leaching procedure (TCLP) was one of two testing methods. In Table 7.1 selected results of their study is shown.

Table 7.1 Worst-case leaching results of metals used in solders [13]

	TCLP leachate result [mg/dm^3]	Solder share [wt%]	Metal share [wt%]	TCLP extraction solution volume [dm^3]
Pb in 63Sn–37Pb	157	1.7	37	2
Ag in 95.5Sn–4Ag–0.5Cu	0.02	1.5	4	2
Sn in 63Sn–37Pb	0.024	1.7	63	2
Bi in 42Sn–57Bi–1Ag	21.1	1.8	57	2

From Table 7.1, Fig. 7.1 below can be derived. There are many measures by which toxicity is expressed.

One of them is the No-Observed-Adverse-Effect-Limit (NOAEL). NOAEL values for Pb, Ag, Sn and Bi are around 0.0035, 0.006, 0.25, and 1,000 mg/kg/day, respectively [10]. The relative leachabilities in Fig. 7.1 refer to kg metal leached per kg metal contained in the solder. Figure 7.1 shows that Pb is more toxic than Ag, Sn, and Bi and also more prone to leach from a larger printed board assembly under extreme conditions. These two facts are important from a conservative hazard perspective as the potential of Pb to cause harm is clearly larger than for the others. Now, the interesting issue is whether any solder metal has been, is, or is thought to be a problem in *controlled* landfills, where most e-waste will end up.

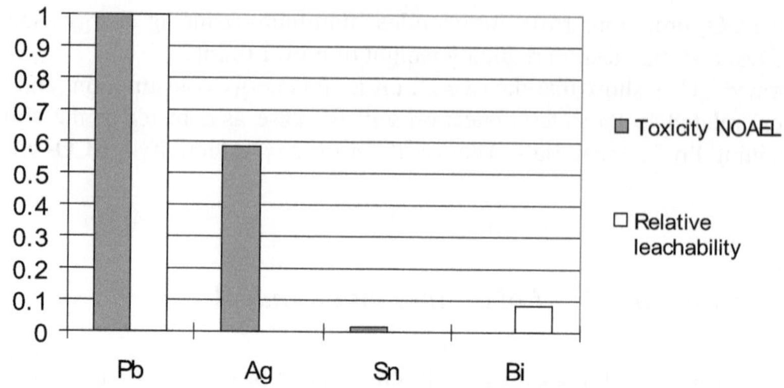

Fig. 7.1 Relative toxicity and leachability for selected metals used in solders and adhesives

LCA is strongly linked to risk assessment [8, 14]. The global associated health and environmental consequences of Pb solders have so far not been proven higher than the ones that will come from Pb-free.

Holistic LCA focusing on all life cycle stages is not misleading, but determineation of the risk associated with the Pb contained in e-waste is controversial.

Suganuma argued that while removal of Pb from e-waste is technically possible, it would create "problems in our social systems" [15].

The conflict, is on one hand, the proven resource losses (dispersion) of Sn and Ag, supported by facts as a very small share of electronic products is recycled, and on the other hand the accumulation of Pb in landfills, which could imply future human health risks.

Would then recycling of Pb-based electronics be the optimal solution to the risk problem associated with Pb? Most authors agree that the Pb-free electronics need to be recycled anyway as the legalization will require it and due to the resource "loss".

In Japan the LIME model was developed among other things to close the gap between risk analysis and LCA. Risk analysis determines the probability of harm from a substance exposed at certain concentration in which is often a local risk situation.

LIME attempts to translate the inventory flows into estimates of environmental damage costs. Without specific impact analysis, the risk of some negative or positive environmental consequences will be impossible to determine.

The toxic potential indicator (Sect. 3.5.2) and the air and water toxicity (Sect. 5.3) proxies are very conservative approaches alike, but can obviously advise where risk analyses are in most demand. The LCI need to be detailed enough as to provide risk analysts with detailed information on the chemical form and location of the toxic emissions. It is my hope that this writing can bring the discussion further on what it takes to reach consensus on whether Pb-free solders are more environmentally compatible than Pb solders.

Here follow embryos of broader theories, identified in connection to the present research, which likely will emerge:

1. A theory on the interlinked global metal production system in order to differentiate metals from each other from LCI and LCA perspectives. Reuter et al. [16] and Nakamura et al. [17] have laid the first building blocks for this theory
2. A theory for identification of marginal consumers in CLCA.
3. A unification of the understanding of damage assessment and integration to single monetary score of environmental cost.
4. Historical measurements of landfill leachate as functions of landfill waste composition.

So far the statistical sciences have not been "pushed forward" by LCI which merely used existing theories.

Obviously, for interconnection materials LCA is only a screening tool which determines which further flows should be considered for a refined risk assessment. Returning to the hypotheses presented in the preface of the present book:

1. The phasing out of Pb from the electronics industry (mainly interconnection materials) is preferable from *different environmental cost perspectives.*
2. Pb emissions will decrease (*in the same way as they did when Pb was phased out as additive to petrol*) if Pb is phased out from the electronics industry.

Based upon laboratory estimations of worst-case landfill Pb emissions, it has been shown for specific LCAs of solder pastes that the LIME environmental cost will decrease. It is an open question what field measurements and risk assessments would imply regarding the environmental cost caused by the shift to Pb-free electronics. The second hypothesis could not be falsified here by technology LCAs.

References

1. Brady TA, O'Connell S, Tanskanen P et al (2003) Material composition profiles of select IT components, a design for environment project with the high density packaging user group (HDPUG). Proc IEEE Int Symp Electron Environ, 19–22 May, Boston, MA, pp 125–130
2. Turbini LJ, Munie GC, Bernier D et al (2001) Examining the environmental impact of lead-free soldering alternatives. IEEE Trans Electron Packag Manuf 24:4–9
3. Müller J, Griese H, Schischke K (2005) Transition to lead-free soldering – a great change for a better understanding of materials and processes and green electronics. Proc Int Conf Asian Green Electron, 15–18 March, Shanghai, China, pp 33–36
4. Müller J, Griese H, Reichl H (1999) Reduced environmental impacts by lead free electronic assemblies? Proc Int Summit Lead-free Electron Assemblies IPCWorks '99, 26–28 Oct, Minneapolis, MN
5. Niu X, Li Y (2007) Treatment of waste printed wire boards in electronic waste for safe disposal. J Hazard Mater 145:410–416
6. Wadanambi L, Dubey B, Townsend T (2008) The leaching of lead from lead-based paint in landfill environments. J Hazard Mater 157:194–200

7. Jang YC, Townsend TG (2003) Leaching of lead from computer printed wire boards and cathode ray tubes by municipal solid waste landfill leachates. Environ Sci Technol 37:4778–4784
8. Pettersen P, Hertwich EG (2008) Life-cycle inventory procedures for long-term release of metals. Environ Sci Technol 42:4639–4647
9. Sleeswijk AW, van Oers LFCM, Guinée JB et al (2008) Normalisation in product life cycle assessment: an LCA of the global and European economic systems in the year 2000. Sci Tot Environ 390:227–240
10. Andrae ASG, Itsubo N, Yamaguchi H et al (2008) Life cycle assessment of high-temperature Japanese conductive adhesives. Env Sci Tech 42:3084–3089
11. Halada K (1992) Ecomaterials. Bull Jpn Inst Metals 31:505–512
12. Halada K (2003) Progress of ecomaterials toward a sustainable society. Curr Opin Solid State Mater Sci 7:209–216
13. Townsend T, Musson S, Dubey B et al (2008) Leachability of printed wire boards containing leaded and lead-free solder. J Environ Manag 88:926–931
14. Matthews HS, Lave L, MacLean H (2002) Life cycle impact assessment: a challenge for risk analysts. Risk Anal 22:853–860
15. Sugnuma K (2003) Lead-free soldering in electronics: science, technology and environmental impact. CRC Press
16. Reuter MA, Verhoef EV (2004) A dynamic model for the assessment of the replacement of lead in solders. J Electron Mater 33:1567–1580
17. Nakamura S, Murakami S, Nakajima K et al (2008) Hybrid input–output approach to metal production and its application to the introduction of lead-free solders. Environ Sci Technol 42:3843–3848

Chapter 8
Conclusions

Abstract This chapter includes the conclusions that can be drawn from the new knowledge generated. Attributional LCA and consequential LCA are complementary but do not always lead to different overall conclusions. As expressed by the LIME method, without measurements of end-of-life emissions, the Sn and Ag consumptions are most important for the negative consequential environmental cost of Sn–Ag–Cu solders introduced instead of Sn–Pb solders. As soon as *estimates* of landfill Pb emissions are introduced the shift from Pb solders seems advantageous seen from a LIME environmental cost perspective. Precious metals such as Pd have a high environmental cost compared to Sn.

The hope of the present book is to bring comprehensiveness to the field of LCA of interconnection materials. The phenomenon of interest is the possible overall ecological compatibility of certain Pb-free solders, as compared to the conventional Pb solders. In the introduction two main hypotheses were addressed.

The first propose that the environmental cost can be estimated using LCA and the second that consequential LCA is mature enough to model changes from Pb to Pb-free. These broad theses were divided into eight testable hypotheses presented in Sect. 3.5.4. In LCA modeling, LIME is an LCIA method which is strong in quantifying environmental cost of global impacts, but needs more sophistication of the LCI to handle local impacts. The CLCA is currently a semi-quantitative discussion approach still in its infancy.

The following conclusions can be drawn based on the performed work:

- Attributional LCA (ALCA) and consequential LCA (CLCA) complement one another in providing knowledge about the possible consequences of the shift in solder paste composition.
- According to the ALCA, the shift from solder paste 62Sn–36Pb–2Ag to 95.5Sn–3.8Ag–0.7Cu is likely to result in reduced Pb emissions and increased CO_2 emissions.
- However, the CLCA (system expansion) of the 62Sn–36Pb–2Ag system reveals that the effect, other than reduced Pb emissions from the solder life cycle, could be increased Pb use in batteries and other products.

- This could lead to more Pb emissions than predicted by the ALCA, but on the other hand lower CO_2 emissions.
- The shift from 62Sn–36Pb–2Ag to 95.5Sn–3.8Ag–0.7Cu solder paste is likely to result in both reduced air toxicity (AT) and water toxicity (WT)
- For AT in a CLCA, a significant increase was detected due to increased Sn production and generation of electricity for solder application, but this increase was offset by a decrease mainly due to avoided incineration of printed board assemblies containing Pb.
- Contrary to the case for global warming and AT, no significant increases in WT could be related to the increased Sn production and generation of electricity for solder application.
- The WT result would be highly dependent on the Pb leachability.
- Identified by CLCA methodology, the alternative use of Pb (storage batteries) will offset the WT reduction related to Pb production to about 50%.
- Compared to 63Sn–37Pb, 95.5Sn–3.8Ag–0.7Cu solder paste is likely to increase globally related environmental impacts. The Japanese LIME weighting method shows that increased Ag and Sn production are the main drivers and that LIME give clear environmental cost results for global impacts.
- For the LIME score *not to be* dominated by landfill Pb emissions, the share of Pb emissions (per Pb in solder) reaching the water environment from landfills has to be around 0.1wt%.
- LCA results based upon DALY damage assessment are dramatically sensitive to the amount of Pb reaching humans via water.
- The solder paste 42Sn–58Bi is more CO_2 advantageous than Zn-free solder pastes and 95.5Sn–3.8Ag–0.7Cu.
- For LIME environmental cost, the plating materials (Sn vs. Au/Ni/Pd) on a QFP component are dominated by Pd driving a several times less mass intensive technology, Au/Ni/Pd, to be worse than Sn.
- Statistically for CO_2 it will be advantageous to shift from 80Ag–ICA to any of 80AlN–ICA, 80Ni–ICA or 80Cu–ICA or a combination of these.
- On the aggregation level of LIME single score, the probability is high that no statistical differences exist in between Ag-free ECAs.
- Scaling-up the board-level LCA profiles to global-level material usage increase the relative differences between CO_2 emissions for ECAs.
- For the resource indicator ore-TMR, 80Ag–ICA shows signs of more "hidden material flows" compared to 10Sn–90Pb and 42Sn–58Bi–ICA. For the last two the existence of a clear difference is in doubt.
- Solder alloy and powder production are insignificant in interconnection LCAs. Energy usage in waste management is also not of principal interest. These life cycle steps and flows can be omitted in future work.

Chapter 9
Looking Ahead

Abstract This chapter proposes the future trends and needs of the research area of LCA of interconnection. The trade-off between global impacts of locally related emissions and the known global impacts should be studied further. Measurements of landfill emissions are needed to find a possible correlation with increasing e-waste content in the landfills. Backcasting CLCI could be worthwhile as the historic time series can be used and the future data estimated. The ecological effects of using nanotechnologies in interconnection are not clear but important to understand before a wider introduction of nanomaterials. For important technologies LCAs need to be performed continually, and not on a one-time basis.

For LIME it has earlier been shown, in a scenario for Japan, that the toxicity of Pb was most important, leading to the overall environmental superiority of Pb-free solders. The next step would be to make a trade-off between the global impact of the locally related impacts, such as those originating from the use of toxins, and the known global impacts. Here, landfills need to be examined regarding both content and emissions. Furthermore, the existence of industrial Pb aerosols could make way for updated LCIA indices.

Measurement of emissions should be set up as the direct and chief aim of the future development of LCIA of interconnection materials. Obviously the leaching behavior of several metals needs further attention as well as the development of DALY-factors. Ce, La, Sn, and Bi are not priorities for toxicity assessment, but could be studied at a later stage.

As appropriate data are quite scarce in certain CLCIs, historical time series data could be explored to find out how large the data needs are for the econometric parts of the CLCI modeling.

42Sn–58Bi materials have shown tremendous ecological properties, at least when evaluated by LCA. However, 42Sn–58Bi is not intended for the same application as, e.g., 95.5Sn–3.8Ag–0.7Cu. Meaning, low-temperature solders should only be compared to each other and likewise high-temperature ones. Globally focused CLCAs of solders need to be improved in a number of areas: (1) the design of quantitative partial economic equilibrium models for the electricity, Sn Ag, and Bi markets as well as the corresponding scrap markets for the metals, (2) identification of marginal consumers for Sn, Ag, Bi and other solder metals, (3) marginal data for primary metal production need to be collected, and (4) system expansion including Au- and Ag-containing printed wiring board surface finishes could be worthwhile.

LCA is ideally a rather quick but still powerful screening tool; LCAs should be performed continually in society to detect trends and develop the methodology. LCAs are not a one-time effort but need continuous updates if they are to support policy makers.

Software tools such as SimaPro 7 should incorporate the possibility to include the uncertainty of the impact and weighting steps. Moreover the software programs, by which the practitioner can perform contribution/dominance, uncertainty, and sensitivity analyses, needs to be developed to also include *t*-test probability analysis. Undergraduate education, sometimes having somewhat limited economic resources, would likely benefit from using EMLCA and CMLCA.

It is important to validate the LCA really has determined whether the environmental benefits using a material technology outweigh its eventual negative environmental effects. Ultimately it is desirable to give the decision makers a result that takes into account the main environmental effects of the decision. This can be done by checking the availability of primary data through the life cycles and applying new modeling methods, especially uncertainty analysis and dynamic modeling.

Section 6.3 suggests that in order to decrease the social cost of the Ag–ICA the recycling of electronic products using it should be increased, especially if Ag–ECAs will be the main replacement for solder paste. The interconnection material component of an electronic product is likely not the major environmental issue at hand; however, it is important that all newly developed materials contribute to lowering the overall system load.

In the best of worlds, for each individual unit process, the LCI data must be linked to environmental impact assessments (EIA) in order to assess the actual effects of a system. Clearly such EIAs have not yet been carried out for the most important land areas which the interconnection materials affect. The LIME based LCIA helps identify the most urgent needs for EIAs.

Some kind of mandatory reporting of LCI data, similar to economic data, would be required. Otherwise the lack of scientific environmental data leads to a huge gap between the actual environmental effects and the environmental loadings.

A state-of-the-art uncertainty calculation should be carried out for the total inventory of a large system especially focusing on the uncertainty arising from impact and weighting steps. The operational part in, e.g., SimaPro 7 is obviously not the bottleneck, but rather the type of distribution for different kinds of LCI and LCIA data.

The CLCA should be explored further using cross-disciplinary research teams consisting mainly of econometricians and LCA experts. CLCA studies of solders should focus on the Ag, Sn, and Bi markets.

Not many research fields escape the nanomaterials, nor do solder pastes and conductive adhesives. The concern is that usually no life cycle or risk assessment is done prior to product life cycle phase of the materials. Research has showed that some nanomaterials require significantly more upstream production energy than conventional materials [1]. For solders and the lowered reflow energy caused by the nanomaterials might offset the upstream elevation. Screening LCAs should be performed on nanosolders, but the inherently uncertain end-of-life fate will still be a problem for such simulations. As if this would not be enough challenges, DALY factors are lacking for nanomaterials but that can be explained by their relatively limited spread in society.

Nevertheless, as soon as possible nanotechnologies are introduced to interconnection concepts, the concepts should be evaluated using the latest impact assessment methodologies.

The ore-TMR indicator is actually one part of Halada's resource scarcity indicator (RSI) index. The other is the so-called resource exhaustion acceleration. The RSI and beyond could be applied to interconnection material technologies, especially Au-plated metal spheres in ACAs. Uncertainty ranges need to be developed for the entire framework of ore-TMR, depletion acceleration coefficient, and resource scarcity index.

Reference

1. Khanna V, Bakshi BR, Lee LJ (2008) Carbon nanofiber production: life cycle energy consumption and environmental impact. J Ind Ecol 12:394–410

Index

O

P

Q

R

S

T

MIX
Papier aus verantwortungsvollen Quellen
Paper from responsible sources
FSC® C105338

If you have any concerns about our products,
you can contact us on
ProductSafety@springernature.com

In case Publisher is established outside the EU,
the EU authorized representative is:
Springer Nature Customer Service Center GmbH
Europaplatz 3, 69115 Heidelberg, Germany

Printed by Libri Plureos GmbH
in Hamburg, Germany